Simplified Procedures for Water Examination

AWWA MANUAL M12

Fifth Edition

American Water Works Association

Science and Technology

AWWA unites the drinking water community by developing and distributing authoritative scientific and technological knowledge. Through its members, AWWA develops industry standards for products and processes that advance public health and safety. AWWA also provides quality improvement programs for water and wastewater utilities.

MANUAL OF WATER SUPPLY PRACTICES—M12, Fifth Edition
Simplified Procedures for Water Examination

Project Manager and Technical Editor: Melissa Christensen
Production Editor: Carol Magin Stearns

Library of Congress Cataloging-in-Publication Data has been applied for.

Printed in the United States of America
American Water Works Association
6666 West Quincy Avenue
Denver, CO 80235

ISBN 1-58321-182-9

 Printed on recycled paper

Contents

Figures

This page intentionally blank

Tables

Foreword

This publication is the third revision of the original American Water Works Association (AWWA) Manual M12, *Simplified Procedures for Water Examination Laboratory Manual*. The original manual, published in 1964, was revised in 1975, with a supplement on instrumental methods added in 1978, and a complete revision was done in 1997. The 1997 edition was revised in 2002.

The original goal for the manual—to provide operators with simplified procedures for tests commonly needed for process control in drinking water production—remains the same. Monitoring plant processes through reliable, reproducible analyses enables plant operators to evaluate and optimize those processes to produce the highest possible quality drinking water.

The manual has been rewritten to include new and updated methods, laboratory equipment, and safety procedures. Information on laboratory equipment includes quick colorimetric kit procedures. AWWA does not intend to recommend any one manufacturer, as most kits are available from a variety of manufacturers. This manual contains basic information on quantitative analysis, but beginners in the laboratory can gain helpful training from additional resources, such as short courses offered by state departments of health, AWWA sections, and chemical or equipment suppliers. Such classroom instruction, demonstrations by experienced teachers, and supervised laboratory work provide valuable additions to independent reading and practice.

These are simplified methods, and it is important to note when more sophisticated methods may be required for compliance monitoring. This manual is not designed to replace *Standard Methods for the Examination of Water and Wastewater*. M12 may serve as a tool to acquire laboratory skills that will eventually facilitate the use of *Standard Methods*. Refer to *Standard Methods* for more information on methods and water quality.

The methods included in M12 are based on the presumption of high-quality water of known and relatively constant composition. Consult state health departments or other regulatory agencies that conduct bacteriological and chemical analyses of drinking waters for advice on the applicability of methods to your situation. Some source waters may require the more complicated procedures from *Standard Methods*. Please note the warning section for each procedure to ensure the applicability of the method.

Appendix A contains a limited list of information sources. The mention of manufacturers or trade names for commercial products does not represent or imply the approval or endorsement of AWWA. Appendix B provides information on safe storage of laboratory chemicals as well as a table of incompatible chemicals. A list of the compounds identified in this manual is available in appendix C, and the periodic table of the elements is at the end of the manual.

If you have any comments or questions about this manual, please call the AWWA Volunteer & Technical Support group, (303) 794-7711 ext. 6283, FAX (303) 795-7603, or write to the group at 6666 W. Quincy Ave., Denver, CO 80235.

This page intentionally blank

Acknowledgments

This edition of AWWA Manual M12 was prepared by the AWWA Water Quality Division Water Quality Laboratory Committee. The membership of the committee at the time it approved this manual was as follows:

R.M. Powell (Chair), Pinellas County Utilities, Largo, Fla.
M. Armacost, EAZ/Systems, Evansville, Ind.
S.N. Choudhuri, Utah Health Laboratory, Salt Lake City, Utah
V. Dwyer, Des Moines Water Works, Des Moines, Iowa
B.R. Fisher, Washington Sub. Sanitary Comm., Laurel, Md.
C. Haas Claveau (Staff Advisor), AWWA, Denver, Colo.
L. Henry, American Water Works Service Company, Okawville, Ill.
C.D. Hertz, Philadelphia Suburban Water Company, Bryn Mawr, Pa.
J.E. Hoelscher Jr., Beaver Water District, Fayetteville, Ark.
D. Kimbrough, Castaic Lake Water Agency, Santa Clarita, Calif.
J.H. Navani, Palm Beach County Water Utilities, Delray Beach, Fla.
C.D. Norton, American Water Works Service Company Inc., Belleville, Ill.
F. Oney, City of Tamarac, Tamarac, Fla.
C.A. Owen, City of Tampa Water Department, Tampa, Fla.
R.M. Saul, Norfolk Utilities Department, Norfolk, Va.
M. Stasiak, Los Angeles Department of Water & Power, Sun Valley, Calif.
I.H. Suffet (Division Liaison), University of California Los Angeles, Los Angeles, Calif.
S.L. Williams, Newport News Waterworks, Yorktown, Va.
C. Wong, San Francisco Water Department, Millbrae, Calif.

The committee wishes to thank the following authors who contributed to this manual: Chris Owen, Sherry Williams, Charles D. Hertz, Jaya Navani, Richard M. Saul, Vincent Dwyer, and Robert M. Powell. Clare Haas Claveau of the AWWA staff deserves special thanks for her advice, support, and assistance to the committee.

This page intentionally blank

Icon Key

The following icons appear on the first page of each procedure to indicate the laboratory equipment necessary to perform that procedure. When a choice of procedures is available, for example a titration setup or a color comparison kit, both icons will indicate those choices.

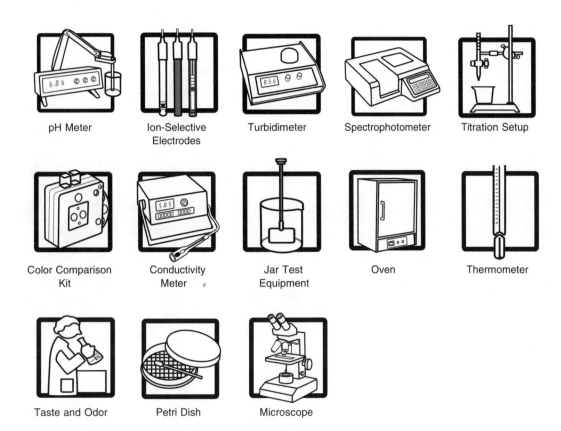

pH Meter

Ion-Selective Electrodes

Turbidimeter

Spectrophotometer

Titration Setup

Color Comparison Kit

Conductivity Meter

Jar Test Equipment

Oven

Thermometer

Taste and Odor

Petri Dish

Microscope

This page intentionally blank

Chapter **1**

Water Quality Laboratory

INTRODUCTION

Reliable laboratory quality testing forms the basis for water quality control. Laboratory test results make it possible for operators to evaluate and optimize plant performance in the following ways:

- Results from water testing help the operator to know and to document water quality conditions throughout the plant.

- Laboratory analyses form a basis for selecting operational procedures and chemical treatment and then for evaluating the effectiveness of changes to the system.

- Logging data and reviewing trends provide the opportunity to identify potential problems before they affect water quality.

- Reliable water quality testing forms the basis for regulatory compliance and ensures the best possible quality drinking water for the community.

LABORATORY PROCEDURES

Temperature and humidity must be controlled in the laboratory. Temperature changes affect the reaction rates of certain test procedures, and high humidity adversely affects the storage life of reagents and media. Storage in direct sunlight contributes to rapid deterioration of some reagents and most media. Unopened bottles of media and chemical reagents should be stored in dark, cool, dry locations, such as storage cabinets, separate from opened bottles. Inverting bottles and storing them cap-side down reduces caking. Consider storing opened bottles of media in a desiccator.

Laboratory facilities need enough counter space to adequately separate bench instrumentation and avoid constantly moving unused equipment to perform testing. Depending on the types of tests conducted, major pieces of equipment may require additional counter or floor space. Laboratory design should include tests currently conducted and accommodate future tests. Good overall laboratory design includes provisions for a vacuum line and natural gas lines, and areas for refrigerators, incubators, ovens, and autoclaves.

Maintaining a clean and efficient work environment is important to sound laboratory procedure. Equipment and supplies should be labeled and allocated a space in the laboratory. Chemicals should be kept in suitable containers and clearly labeled. The laboratory should be kept as clean and dust free as possible, and equipment should be protected from moisture and harmful fumes. There should be adequate lighting for color comparisons and ample facilities for lubricating, washing, and drying laboratory equipment, and for other laboratory activities. Testing equipment and supplies should be handled with care and respect. The laboratory should never be used as a lunchroom, kitchen, or miscellaneous storage facility.

Incorrect sampling or laboratory procedures in water testing are as bad as, or worse than, no testing at all. Inaccurate information obtained from contaminated samples or that result from incorrect procedures may lead to incorrect water treatment decisions. Accurate analysis depends on sampling procedures, careful maintenance and calibration of equipment, and attention to laboratory procedures.

SAMPLING TECHNIQUES

Water sampling seems to be a simple procedure. However, any time a small sample is withdrawn from a larger body of water, there is a potential for error. Most major errors in water quality analysis are caused by poor sampling. When sampling is inadequate or not representative, water treatment decisions will be based on inaccurate information. It is essential to be particularly careful to obtain representative samples from the water source, use the proper sampling technique for that source, and preserve samples until they are analyzed.

The location and frequency of sampling will be governed by the type of source water available. Samples may be obtained either as *grab* or *composite* samples.

Grab samples are most useful when an analysis is needed of the water characteristics at a single point in time. This is particularly important for water quality indicators that may change over time. These include dissolved gases, residual chlorine, pH, temperature, and coliform bacteria. Grab samples are most useful when the water has characteristics that remain relatively constant and flow is not continuous.

Where water characteristics change continuously, such as in streams, composite sampling is most effective if done continually. However, in many cases this is impossible. In composite sampling, water samples are collected at regular intervals, perhaps hourly or every 2 hr. Portions of these samples are then combined and analyzed at the end of 24 hr. The size of the portion to be mixed with other sample portions is proportionate to the water flow at the time the sample was collected.

Representative Sampling

Because a large body of water is not uniform in quality throughout, the best procedure is to take samples from many sites, analyze them separately, and evaluate data as a group to determine the quality of that body. This procedure is more likely to be representative of the entire body than a single sample taken at one point. The more points from which samples are withdrawn, the more representative of actual water quality the overall data set will be. Good judgment should be used to select a sampling method, and several factors should be considered including

- the character of the laboratory examinations to be made
- how test results will be used (the objectives of the analysis)
- the nature of the water and variations in its characteristics over time
- the variation of the flow rate over a sampling period

Wells. Pumped water is usually the only source for well sampling. The mix in well water is generally good, and composite sampling is rarely needed. Pumps and casings can contribute to sample contamination, and the sample may not be representative, particularly if a pump is not used often. Changes in water quality, i.e., arsenic, nitrate, and some organic contaminants may also occur in certain formations with the length of time a well is pumped and the well drawdown.

Lakes and reservoirs. Representative sampling of lakes and reservoirs is often difficult because of changes in water composition at various depths and temperatures. For this reason, a single sample represents only one area of the lake. Depending on the size of the impoundment, several samples must be collected at various areas and depths. The larger the impoundment, the more samples are needed if the objective is to fully characterize the system.

Rivers. Small- or medium-sized streams are usually more easily sampled than larger rivers because it is possible to find a place where the water is uniform or well mixed. If such a location cannot be found, it is advisable to collect samples from several locations. This is often necessary in larger streams and rivers.

In-plant sampling. Most sampling is done at the water treatment plant. Large plants may have continuous sampling provisions, but such systems may be too expensive for small plants. A sample tap must be located properly to ensure that samples are representative. A tap installed in an area that does not provide a representative sample is useless for obtaining accurate information.

Distribution system sampling. Obtaining representative samples throughout the distribution system is probably the best way to determine water quality throughout the system. Sampling sites should be selected to trace the course of finished water through mains and then through the major arteries of the system. A short, corrosion-resistant connection to the main is an ideal sampling site. If special sample taps are not available, samples may be collected from customers' faucets. Fire hydrants are generally not satisfactory sampling sites because the flow is erratic and they are not used often; therefore, samples become contaminated from corrosion or sediment.

Sampling Devices

Sampling may be done either automatically or manually. Automatic samplers save time in the sampling process, but they are expensive and may require frequent maintenance. Staff must be aware of these possibilities and check regularly for possible problems to prevent sampler malfunctions. Dippers, weighted bottles, hand-operated pumps, and similar equipment are classified as manual sampling devices. A weighted bottle or other collection container is used to collect depth samples. Depth samplers are designed to be lowered in the open position with a mechanism that allows for a valve or stopper to be closed when the container is filled.

Sampling Techniques

Surface sampling. Surface water may be sampled easily by using the following procedure:

1. Grasp the sample container at the base with one hand and submerge it, mouth down.

2. Position the mouth of the container into the current and away from the collector's hand.

3. If there is no current, create an artificial one by moving the container in the direction it is pointed.

4. Tip the bottle up to allow air to escape and water to enter the container.

5. Remove the bottle from the water.

6. Stopper the container tightly.

7. Label the container with date, time, type of sample, location, name of collector, and water temperature. Fill out chain of custody form, if applicable.

If the sample must be taken from a bridge, walkway, or other structure above the water surface, first place the container in a frame that has enough weight to submerge the container. Attach a nylon or other nonrotting rope to the frame and remove the stopper. Lower the frame and container to just above the water level, facing the current. Swing it downstream slightly and drop it into the water. Pull the container upstream and out of the water. Stopper and label it.

Water tap sampling. When sampling from water main connections, flush sediment and possible contaminants from the line before collecting the sample. Never take samples from leaking or corroded taps, from taps surrounded by dense vegetation, from public drinking fountains, restrooms, or from taps with aerators. To collect a sample:

1. Use a container that is clean and free from dust or other contaminants.

2. Clean the outer parts of the tap to prevent debris from falling into the container. Disinfect the tap if appropriate.

3. Flush the line by partially opening the tap and allowing the water in the line to run freely until it reaches the known temperature of distribution mains in the area. This may take 4 or 5 min. (Do not turn the water on full to save time. This may disturb incrustations and sediment in the line and contaminate the sample.)

4. Place the container close to the tap but do not allow it to touch the connection.

5. Collect the sample.

6. Stopper and label the container. The label should show date, time, type of sample, location, name of collector, and water temperature. Fill out chain of custody form, if applicable.

Sample Preservation

Samples should be analyzed as quickly as possible after sampling. Some analyses, such as those for temperature, pH, chlorine, and sulfite, must be done immediately. Other analyses may be done as long as 6 months after the sample is taken (if the sample is properly preserved). However, the sooner the analysis is done, the more quickly any problems related to sample quality, test method, or water quality can be resolved.

USING THE METRIC SYSTEM

In a laboratory, chemicals are weighed in grams (g) and milligrams (mg), and liquids are measured in liters (L) and milliliters (mL). Temperature is measured in degrees Celsius or centigrade ($°C$) rather than in Fahrenheit ($°F$). Length is usually measured in meters (m) and centimeters (cm) rather than in inches (in.), feet (ft), or yards (yd). The metric system, like the decimal system, is based on units of 10. If a measurement is less than 1 meter, liter, or gram, a prefix signifies how much less. For example, the following are some common prefixes used in the laboratory: a centimeter equals one hundredth of a meter, or expressed in decimals, it is 0.01 meter; a millimeter (mm) is one thousandth or 0.001 meter; and a micrometer (μm) is one millionth of a meter or 0.000001 meter. Measurements larger than meters, liters, or grams also use prefixes to signify the multiplier. For example, a kilometer (km) is 1,000 meters.

Conversion Factors

Standard reports are made in milligrams per liter (mg/L). One L of water weighs close to 1 million mg, so 1 mg of a substance in 1 L of water represents 1 part per million (ppm). The unit ppm used in water analysis always means by weight, never by volume. For example, 10 ppm hardness of calcium carbonate ($CaCO_3$) means 10 mg of $CaCO_3$ per 1 million mg water or 10 pounds of $CaCO_3$ per 1 million pounds water.

Table 1-1 provides a set of conversion factors useful in converting milligrams per liter to grains per gallon or milligrams per liter to pounds per 1,000 gallons.

WORKING WITH CHEMICALS

All prepared solutions should be of the best available quality. Chemicals labeled "ACS grade," "primary standard grade," or "analytical reagent grade" and dyes certified by the Biological Stain Commission yield the best results. In the methods in this manual, two names are sometimes given for the same chemical (such as potassium dihydrogen phosphate and potassium monobasic phosphate, both KH_2PO_4). These refer to the same substance and the chemical formula follows the name.

A compound is a substance composed of two or more elements. As a rule, all chemical compounds are divided into organic compounds, those that contain carbon (C), and inorganic compounds that have no carbon. However, there are a few exceptions. These include the inorganic compounds of carbon dioxide (CO_2), carbon monoxide (CO), bicarbonate (HCO_3^-), and carbonate (CO_3^{2-}). Table 1-2 provides a list of compounds commonly used in water treatment. For a more extensive listing of the compounds in this manual, see appendix C.

Table 1-1   Conversion factors

	ppm or mg/L	Equivalent			
		gr/ US gal	lb/ 1,000 US gal	gr/ Imp gal	lb/ 1,000 Imp gal
1 part per million	1	0.0583	0.00834	0.0700	0.100
1 grain per US gallon	17.1	1	0.143	1.20	0.172
1 pound per 1,000 US gallon	120	7	1	8.41	1.20
1 grain per Imperial gallon	14.3	0.833	0.119	1	0.143
1 pound per 1,000 Imperial gallon	99.8	5.83	0.833	7	1

Table 1-2   Chemicals commonly used for water treatment

Chemical Name	Formula
Acetic acid	CH_3COOH
Aluminum sulfate	$Al_2(SO_4)_3 \cdot 18H_2O$
Ammonium hydroxide	NH_4OH
Calcium carbonate	$CaCO_3$
Copper sulfate	$CuSO_4$
Ferric chloride	$FeCl_3$
Nitric acid	HNO_3
Phenylarsine oxide	C_6H_5AsO
Sodium bicarbonate	$NaHCO_3$
Sodium hydroxide	$NaOH$
Sulfuric acid	H_2SO_4

Many compounds can be created from the same two or three elements. This formula should be checked with the formula on the container label to confirm that the right chemical is used. A chemical symbol is often used in the laboratory as a symbol for the names of the elements. For example, in the periodic table of the elements (inside back cover), calcium is expressed as Ca and carbon as C. Because chlorine's symbol is Cl and copper's symbol is Cu, mistakes can easily be made in identifying chemicals in the laboratory. It is wise to double-check the name of the chemical and its formula to make sure the correct chemical is used. Chemicals behave very differently in solutions, and a mistake in the solution can result in errors and even hazards.

All bottles that contain standard solutions should be labeled with the name of the solution, its concentration, date of preparation, name of the person who prepared the solution, and, in some cases, the procedure for which the solution is intended.

The word *normal* (abbreviated N) in front of a reagent's name indicates the concentration or strength. The word *standard* is also used, but means merely that the concentration is exactly known. The concentration of a standard solution is sometimes referred to as its *normality*. Thus, the normality of the standard acid used to determine the alkalinity of water is 0.02. This can be written $0.02N$, $N/50$, or $1/50N$, all of which have the same meaning. The important thing to remember about *normal* solutions is that, for example, 100 mL of a $0.02N$ sodium hydroxide ($NaOH$) solution will exactly neutralize 100 mL of a $0.02N$ sulfuric acid (H_2SO_4) solution. A $0.1N$ sulfuric acid solution, on the other hand, is a different and stronger concentration of the acid.

With the high-quality reagents available from chemical supply sources, solutions can be prepared easily for analyses in the laboratory by adding distilled or deionized (reagent grade) water. However, they must be measured and standardized carefully before they are used in a laboratory procedure. The standardization of each solution is described in *Standard Methods for the Examination of Water and Wastewater*. Preparing reagent solutions requires more time and effort, but is less expensive than using prepared solutions. Many standard solutions can be purchased from reliable supply houses, and their use is recommended for individuals who may not have the experience, time, or equipment to prepare them.

Standard Calibration Curves

Regularly standardizing calibration curves provides an opportunity to identify any problems with the procedure, such as deteriorated reagents, need for instrument maintenance, dirty glassware, improper technique, or deterioration of reagent-grade water. Checking calibration curves frequently is an important aspect of quality assurance. To prepare a standard for colorimetric analysis:

1. Weigh dry reagents on a calibrated balance, or measure a solution with a calibrated micropipet, or use a commercially prepared standard.

2. Using distilled or deionized (reagent grade) water and a standard solution of the compound, prepare at least five standards that encompass the lower and upper concentration range and three equally spaced concentrations in between, plus a reagent blank. Five points cannot be used for some parameters, such as pH (2 or 3 points) and turbidity (1 point). Figure 1-1 shows how to prepare decimal dilutions from a stock solution of 10,000 mg/L.

3. Add the reagents in the required sequence to develop the color in each standard. Follow the exact steps performed in the sample analysis. Pay

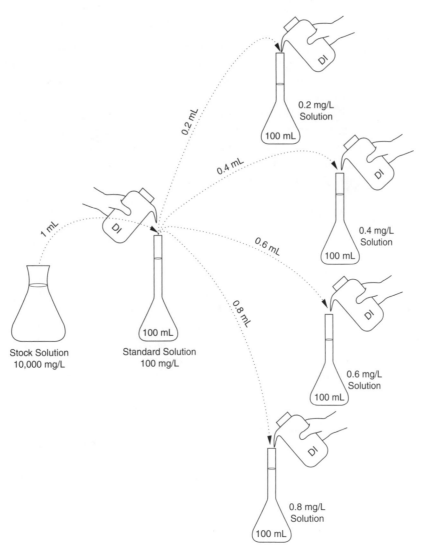

Courtesy of Fort Collins Water Treatment Plant, Fort Collins, Colo.

Figure 1-1   Making solutions for the standard curve

special attention to proper spacing and timing of the standards and samples when color development time is important.

4. Pour an adequate volume of each developed standard into separate, clean, and matched sample tubes, vials, or cuvettes that fit into your colorimetric instrument. Begin with the lowest concentration and move to the highest. Rinse between samples with reagent-grade water, then with a small portion of the standard or sample when reusing cuvettes or vials. The colorimetric instrument should have enough time to warm up and be adjusted to the correct wavelength, or have the correct color comparator or filter in the designated slot.

5. Zero the instrument with distilled water or a reagent blank prepared exactly as the standards and samples. Place the sample in the instrument, making sure the sample compartment is tightly closed. Note the absorbance or percent transmittance readings of each standard. For analog instruments, adopt the correct eye position while reading the results.

6. Plot the percent transmittance on the logarithmic scale versus the concentration on the linear scale of semilogarithmic graph paper.

7. Draw a smooth curve to connect the points (as shown in Figure 1-2). A straight line that starts from the zero point on the graph indicates an ideal color system for colorimetric use. Although calibration curves tend to deviate from straight lines at high and sometimes low concentrations (i.e., NO_3^-) of the compound, other causes of deviation may originate in stray light because of light leaks, optical imperfections, or improper optical alignment or maintenance. The information in Table 1-3 allows conversion from percent transmittance to absorbance readings if semilogarithmic graph paper is unavailable.

SAFETY PROCEDURES

Laboratory safety is important. There are many hazardous materials in the water laboratory, and everyone in the laboratory must remain alert and careful to avoid danger. For information on safe storage of chemicals, see appendix B. Take an extra moment or two to follow a safe practice rather than to risk injury. Be especially careful at the end of a shift or when tired, because most injuries occur at those times.

For specific safety questions, refer to state General Industrial Safety Orders or to Occupational Safety and Health Act (OSHA) regulations. Information on setting up a program to manage hazardous chemicals may be found in US Environmental Protection Agency and OSHA publications. Chemical suppliers provide Material Safety Data Sheets (MSDS) for chemicals used in the laboratory and in water treatment. The MSDS must be readily available to all plant personnel and updated as chemicals are added or deleted (Figure 1-3). Several chemical suppliers also offer smaller quantities of reagents to minimize the quantity of hazardous chemicals on site and ensure freshness.

Safety and right-to-know information must be clearly posted for all plant employees. Figures 1-4 and 1-5 show how a medium-size water treatment plant provides safety and hazardous materials information to plant staff.

First aid equipment should be readily available for laboratory use. Figure 1-6 shows a first aid station near the laboratory. It includes a first aid kit, stocked weekly, and a 5-min minimum emergency oxygen supply.

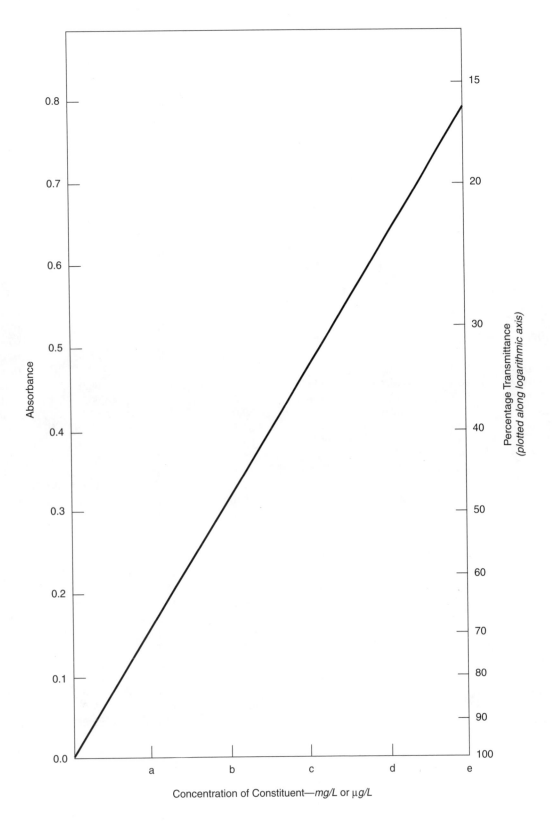

Figure 1-2   Typical photometric calibration curve

Table 1-3   Conversion from percent transmittance to absorbance

Transmit-tance percent	Absorbance	Trans-mittance percent	Absorbance	Transmit-tance percent	Absorbance	Transmit-tance percent	Absorbance
1.0	2.000	26.0	.585	51.0	.292	76.0	.119
1.5	1.824	26.5	.577	51.5	.288	76.5	.116
2.0	1.699	27.0	.569	52.0	.284	77.0	.114
2.5	1.602	27.5	.561	52.5	.280	77.5	.111
3.0	1.523	28.0	.553	53.0	.276	78.0	.108
3.5	1.456	28.5	.545	53.5	.272	78.5	.105
4.0	1.398	29.0	.538	54.0	.268	79.0	.102
4.5	1.347	29.5	.530	54.5	.264	79.5	.100
5.0	1.301	30.0	.523	55.0	.260	80.0	.097
5.5	1.260	30.5	.516	55.5	.256	80.5	.094
6.0	1.222	31.0	.509	56.0	.252	81.0	.092
6.5	1.187	31.5	.502	56.5	.248	81.5	.089
7.0	1.155	32.0	.495	57.0	.244	82.0	.086
7.5	1.126	32.5	.488	57.5	.240	82.5	.084
8.0	1.097	33.0	.482	58.0	.237	83.0	.081
8.5	1.071	33.5	.475	58.5	.233	83.5	.078
9.0	1.046	34.0	.469	59.0	.229	84.0	.076
9.5	1.022	34.5	.462	59.5	.226	84.5	.073
10.0	1.000	35.0	.456	60.0	.222	85.0	.071
10.5	.979	35.5	.450	60.5	.218	85.5	.068
11.0	.959	36.0	.444	61.0	.215	86.0	.066
11.5	.939	36.5	.438	61.5	.211	86.5	.063
12.0	.921	37.0	.432	62.0	.208	87.0	.061
12.5	.903	37.5	.426	62.5	.204	87.5	.058
13.0	.886	38.0	.420	63.0	.201	88.0	.056
13.5	.870	38.5	.414	63.5	.197	88.5	.053
14.0	.854	39.0	.409	64.0	.194	89.0	.051
14.5	.838	39.5	.403	64.5	.191	89.5	.048
15.0	.824	40.0	.398	65.0	.187	90.0	.046
15.5	.810	40.5	.392	65.5	.184	90.5	.043
16.0	.796	41.0	.387	66.0	.181	91.0	.041
16.5	.782	41.5	.382	66.5	.177	91.5	.039
17.0	.770	42.0	.377	67.0	.174	92.0	.036
17.5	.757	42.5	.372	67.5	.171	92.5	.034
18.0	.745	43.0	.367	68.0	.168	93.0	.032
18.5	.733	43.5	.362	68.5	.164	93.5	.029
19.0	.721	44.0	.357	69.0	.161	94.0	.027
19.5	.710	44.5	.352	69.5	.158	94.5	.025
20.0	.699	45.0	.347	70.0	.155	95.0	.022
20.5	.688	45.5	.342	70.5	.152	95.5	.020
21.0	.678	46.0	.337	71.0	.149	96.0	.018
21.5	.668	46.5	.332	71.5	.146	96.5	.016
22.0	.658	47.0	.328	72.0	.143	97.0	.013
22.5	.648	47.5	.323	72.5	.140	97.5	.011
23.0	.638	48.0	.319	73.0	.139	98.0	.009
23.5	.629	48.5	.314	73.5	.134	98.5	.007
24.0	.620	49.0	.310	74.0	.131	99.0	.004
24.5	.611	49.5	.305	74.5	.128	99.5	.002
25.0	.602	50.0	.301	75.0	.125	100.0	.000
25.5	.594	50.5	.297	75.5	.122		

Eyewash and shower stations illustrated in Figure 1-7 should be immediately accessible from the laboratory. Employees who might be exposed to chemical splash should be trained to

- remove protective equipment after activating the eyewash station.

- rinse any chemicals that may come in contact with eyes or skin.

- remove all affected clothing.

- remove contact lenses.

- Hold eyelids open and roll the eyeballs so water flows on all surfaces and in the folds surrounding the eyeballs.

Figure 1-3   Material Safety Data Sheets

Figure 1-4   Plant information and safety center

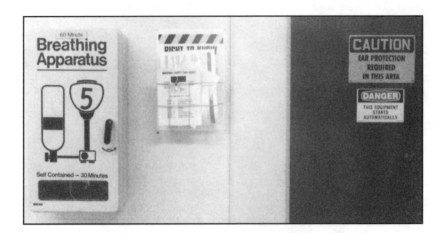

Figure 1-5   Right-to-know information

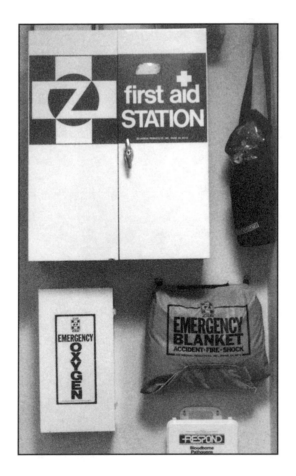

Figure 1-6   First aid station

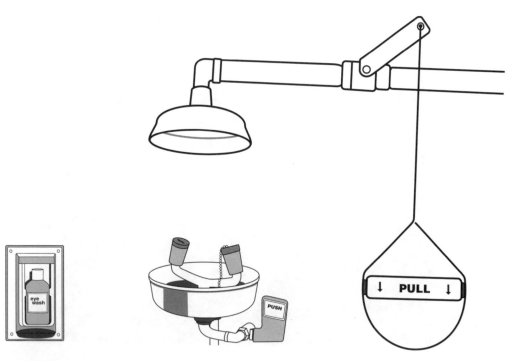

Figure 1-7   Eyewash and shower stations

Corrosive and Toxic Materials

To avoid injury, all chemical reagents should be handled with care both in the form purchased and after dilution. Use extreme caution when handling chemicals marked *poison*, *danger*, *caution*, or *flammable*. These chemicals should be kept in their original containers or in carefully marked containers used only for those solutions. Using a fume hood (shown in Figures 1-8 and 1-9) can prevent injury to laboratory employees by containing and removing hazardous fumes and hazardous chemicals while procedures are being performed. Any danger should be clearly posted on the hood. Most fume hoods pull air past the operator into the hood at about 50 to 100 ft per min as the sash opening is changed. Standards are available to check the air flow against capacity recommendations. Most fume hoods also contain cabinets that are useful for storing hazardous materials, but always guard against storing incompatible chemicals in the same place.

Hydrazine sulfate ($N_2H_4{\cdot}H_2SO_4$), used to prepare the formazin solution suspension for calibrating the nephelometer, is a carcinogen. Laboratories that want to eliminate the hazard of hydrazine sulfate exposure may purchase the stock 4,000 NTU standard suspension from commercial vendors and eliminate the need to purchase, store, and handle hydrazine sulfate.

Safety, automatic, and bulb-operated pipets are recommended for measuring and transferring chemical solutions that can inflict injury by inhalation, absorption, or contact through the mouth, lungs, and skin. **Caution: never pipet any liquids by mouth.**

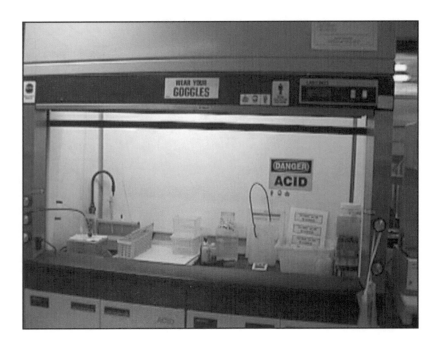

Figure 1-8   Fume hood

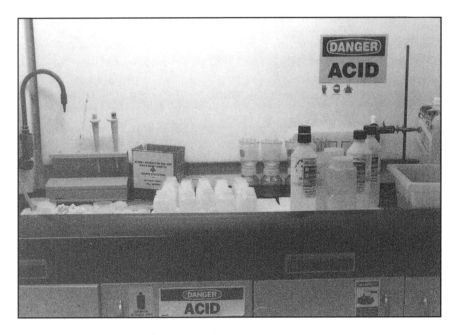

Figure 1-9   Safety procedures using fume hood

Preventing Accidents

Work in the laboratory can be hazardous, but following basic precautions can reduce the risks. Figure 1-10 illustrates essential items for personal safety in the laboratory.

- If possible, do not work alone in the laboratory. If this situation is unavoidable, another person should check in the laboratory regularly in case an accident occurs.

- Wear protective clothing, such as safety glasses, safety shoes, a laboratory coat, and a rubber apron to protect from spills.

- Clean up spills immediately with the proper spill kit.

- Wear insulated gloves or use tongs when handling hot materials. If material might erupt from a container, wear a face shield and goggles.

- Flush the outside of acid bottles with water before opening. Always pour acid into water (never water into acid) to avoid splashing. Never place stoppers or utensils on the counter after use. Always be careful of heat generated. Rinse the outsides of acid containers after use.

- Secure gas cylinders to prevent tipping or rolling. Use a hand truck to move cylinders. Never roll a cylinder by its valve.

- Immediately wash off any chemicals that spatter with large quantities of water. Use vinegar to neutralize bases and baking soda to neutralize acids.

- **Never** eat, drink, smoke, chew gum, or apply cosmetics where laboratory chemicals are present. Always wash hands before doing any of these things.

- Do not smell or taste chemicals. Vent any apparatus that might discharge toxic chemicals, such as vacuum pumps and distillation columns, into local exhaust devices.

- Use a self-contained breathing apparatus (Figure 1-5) when working with chlorine and other toxic substances. Always use a fume hood with adequate air displacement when using hazardous or volatile chemicals. Have the laboratory building ventilation systems and fume hoods checked monthly. Some chemicals are volatile and require use of fume hoods. Some are safe to use without hoods. MSDS sheets and other chemical references are used to determine which chemicals require hoods.

- Neutralize corrosive materials before disposing of them in corrosion-resistant sinks and sewers. Flush with large amounts of water.

- Do not force glass tubing, thermometers, or other glass objects through rubber connections. Wet or lubricate the joints to avoid breakage, and use thick protective gloves and goggles.

- Ensure that properly labeled fire extinguishers and a fire blanket are available at all times (Figure 1-11). Fire extinguishers are classified by the type of fire they will control. Be careful when using a fire extinguisher on small container fires because the force of the spray might knock over the container and spread the fire. Use a fire blanket to smother clothing fires.

- Ensure that a first aid kit is available at all times (Figure 1-6). Space should be available to post proper emergency first aid procedures in plain sight or store a manual in an accessible place (Figure 1-5).

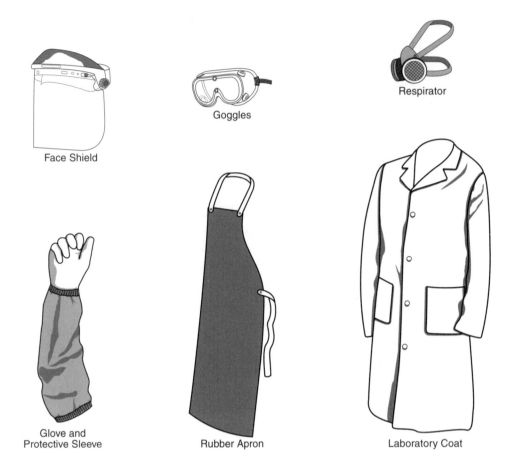

Figure 1-10   Personal safety equipment

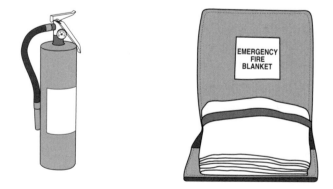

Figure 1-11   Laboratory safety essentials

EQUIPMENT

Laboratory equipment is divided into categories in the following sections to help readers locate equipment. Sections are arranged beginning with glassware, then weighing and measuring equipment, followed by equipment used for steps in a variety of procedures, and finally, test-specific equipment such as pH and conductivity meters.

Glassware

Glassware used in the laboratory should be heat resistant, borosilicate, class A glassware (trade names for class A glassware include Pyrex and Kimax). Bottles, beakers, and stirring rods made of polyethylene are suitable for some operations, but plastic ware does not withstand high temperatures or strong oxidizing solutions. Disposable glassware, such as pipets, test tubes, and bottles, may be used if the laboratory lacks time or equipment necessary for cleaning. Teflon-lined stoppers and caps should be available for some applications.

Pay special attention to ordering and using glassware intended for measuring liquids, because some vessels are designed to contain (TC) or to deliver (TD) the specified amount of fluid. All TC or TD beakers and pipets must have clearly marked volumes with a 2.5 percent tolerance or less.

Beakers. Beakers, as shown in Figure 1-12, are perhaps the most common pieces of laboratory equipment. They are used to mix and measure chemicals. Measurements from a beaker are approximate volumes. Beakers range in size from 1 mL to 4 L.

Flasks. Flasks come in various shapes and sizes and are used to hold and mix reagents. Figure 1-13 illustrates a variety of flasks. Volumetric flasks, squat bottles with long, narrow necks, are used for measuring a specific volume. Their typical capacity ranges from 25 to 2,000 mL, indicated by an etched ring around the neck.

A filled volumetric flask should be viewed at eye level so the front and back sections of the ring around the neck merge into a straight line, and the bottom of the water level (the meniscus) touches this line. In using volumetric glassware, such as flasks, graduates, burets, and pipets, always read the quantity at the bottom of the curve (the meniscus). See Figure 1-18 under buret description for more information on the meniscus.

Volumetric flasks are used to prepare and dilute standard solutions. For example, $0.02N$ acid can be prepared by measuring with a transfer pipet 50 mL of stock $0.1N$ acid solution and placing the acid in a 250-mL volumetric flask. After

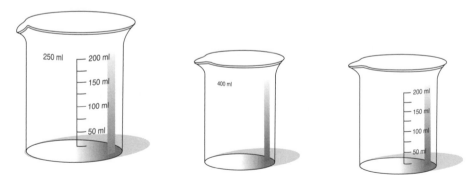

Figure 1-12   Beakers

filling to the mark with distilled water, mix the contents thoroughly by stoppering and inverting the flask 15 times or more. Because volumetric flasks are designed for measuring purposes, the contents should be poured into a clean storage bottle. Erlenmeyer flasks are not used for measuring but are useful for mixing and other laboratory processes.

Graduated cylinders. The workhorse of laboratory glassware is the graduated cylinder, popularly called the *graduate*. See Figure 1-14. Available in sizes as large as several liters, graduates are marked in milliliters, except for the 10-mL size, which is subdivided into fractions of a milliliter, and sizes of 250 mL or larger, which are etched at intervals of 5 or 10 mL.

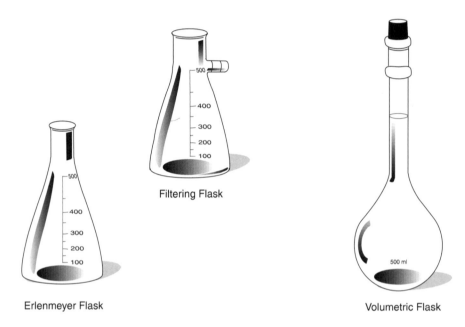

Erlenmeyer Flask

Filtering Flask

Volumetric Flask

Figure 1-13   Varieties of flasks

Figure 1-14   Graduated cylinder

Pipets. Two kinds of pipets are in general use. See Figure 1-15 for the most common pipets in the water process lab. Those with a graduated stem are called *measuring pipets* and can measure any volume up to the designated capacity of the pipet. Those with a single etched ring near the top are called *transfer* or *volumetric pipets*.

One end of the pipet is tapered; the other end is fire polished so it can be easily closed by the pipet bulb. The small tapered end of the pipet is inserted in the bottle, the liquid drawn above the upper etched ring by suction from the pipet bulb (Figure 1-16). The pipet's tip is wiped dry with a clean cloth except during bacteriological examinations. The air is expelled from the bulb just enough to allow the liquid to gradually fall to the desired level. The measured liquid is allowed to flow freely into the receiving container by entirely removing the bulb from the top of the pipet. After the liquid stops draining, the pipet's tip is touched to the inner surface of the receiving container to remove the last drop from the pipet.

When drawing liquid into a pipet, always keep the tip submerged as long as suction is being applied. Avoid vigorous suction during the filling operation because bubbles may form and rise to the surface, where they may take some time to break, making an exact reading difficult. Medicine droppers are convenient to dispense small quantities from a few drops to 1 mL.

Warning: Never use mouth suction.

For accurate work, measure samples with volumetric pipets. When speed is desired at a slight sacrifice in accuracy, carefully measure sample volumes with 100- or 50-mL graduated cylinders for results that are acceptable on an occasional basis. In the range below 50 mL, however, rely on volumetric pipets to measure sample volumes because measuring errors can play a significant role in the final result.

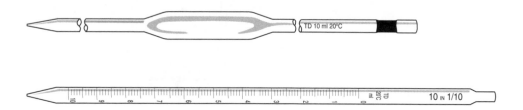

Figure 1-15   Transfer (volumetric) and measuring pipets

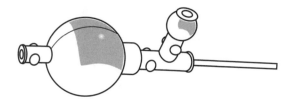

Figure 1-16   Pipet bulb

Burets. A buret is a glass tube graduated over part of its length. The most commonly used sizes are 10, 25, and 50 mL. The graduations in tenths of a milliliter help to estimate a fraction of a tenth or hundredths of a milliliter. Clamp the buret to a stand as shown in Figure 1-17 and fill from the top, usually through a funnel. Allow the excess solution above the zero line to drain to waste through the stopcock before starting a titration.

When reading a buret, take care to read the level of liquid at eye level as in Figure 1-18, not from above or below. A buret that contains liquid shows a curvature in its upper surface, known as the meniscus. Mercury curves upward and water curves down.

Always read a buret twice. The difference between the readings represents the volume of titrant dispensed. Take precautions against air bubbles that arise from the failure of the titrant to wet the buret uniformly through its entire length. Also take care to avoid a slowly leaking stopcock, which allows the titrant to drain as the buret stands idle.

Teflon stopcocks are preferable because glass stopcocks need frequent lubrication to function satisfactorily. For best results with glass stopcocks, apply a lubricant, such as petroleum jelly, as a thin film on the dried surface of the stopcock. Use the

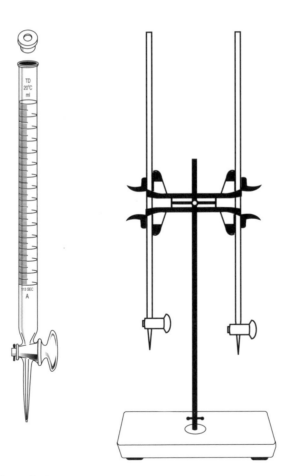

Figure 1-17   Buret, stand, and clamp

lubricant sparingly, because an excess on the stopcock can clog the buret tip and contribute to the formation of air bubbles in the tip.

A buret is specifically designed for dispensing a titrating solution, and individuals who find it difficult to control a pipet may find a buret, or a student buret, a convenient substitute for measuring the accurate volumes of standard solutions needed in colorimetric work.

Automatic pipets and dispensers. These convenient laboratory aids have the advantage of delivering a premeasured quantity of reagent solution quickly. Automatic pipets and micropipets (Figure 1-19) are particularly helpful in the exacting process of preparing standard curves, and must be calibrated on the balance. Dispenser bottle tops (Figure 1-20) save time when running analyses.

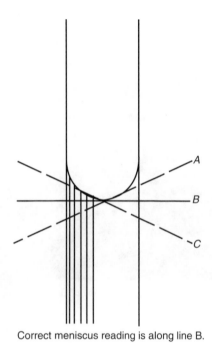

Correct meniscus reading is along line B.

Figure 1-18   Reading of meniscus

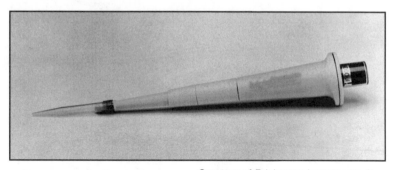

Courtesy of Brinkmann Instruments, Inc.

Figure 1-19   Automatic pipet

Automatic titrators. Improvements have been made in the classical buret with the intent to increase the speed of titrations. These titrators have a buret and quickly transfer the titrant. This type of buret system is available in several forms, two of which are shown in Figures 1-21 and 1-22.

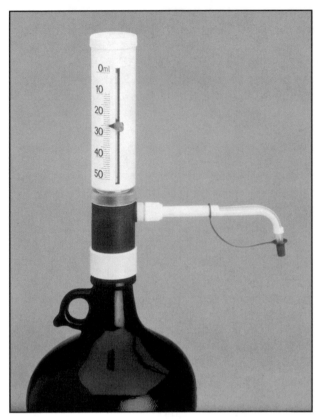

Courtesy of Brinkmann Instruments, Inc.

Figure 1-20   Dispenser bottle top

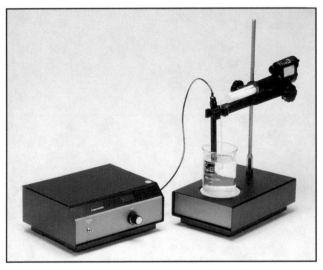

Courtesy of Hach Company

Figure 1-21   Amperometric titrator

Bottles. Bottles commonly used for mixing, storing, and dispensing reagents and samples include glass and plastic sample and reagent bottles, biochemical oxygen demand (BOD) bottles and milk dilution bottles, illustrated in Figure 1-23.

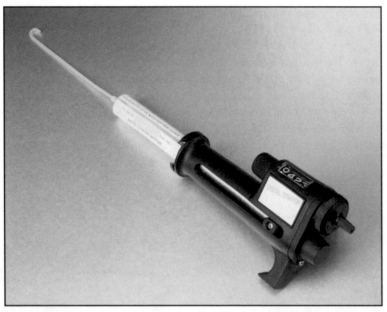

Courtesy of Hach Company

Figure 1-22   Digital titrator

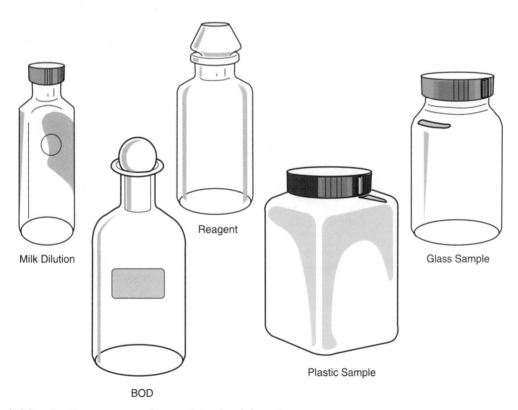

Figure 1-23   Bottles commonly used in the laboratory

Crucibles. Crucibles are glass, porcelain, or metal dishes, shown in Figure 1-24, that are particularly useful for drying and mixing operations in titrations.

Funnels. Funnels are used to pour solutions or transfer chemicals from one container to another. Specialized funnels that separate solids from a mixture (Buchner funnel) or that separate one chemical mixture from another (filter funnel), are available. See Figure 1-25.

Petri dishes. These are shallow, flat glass dishes with low, straight sides and loose covers, illustrated in Figure 1-26. They are commonly used for microbiological cultures. Petri dishes may be made of Pyrex glass that can be sterilized in an autoclave or hot air oven, or of disposable, presterilized plastic.

Test tubes. Test tubes come in many sizes and are stored in racks when in use (see Figure 1-27). They serve as containers for certain chemicals, liquids, or bacteriological media, or they can be used for mixing small quantities of chemicals. Commonly used tubes include the chemical, serological, and bacteriological. The Durham tube is a specialized tube inverted inside a larger tube that contains media

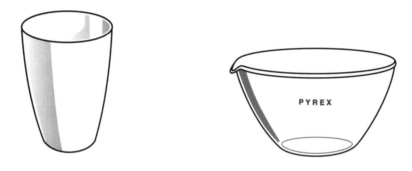

Figure 1-24   Crucibles

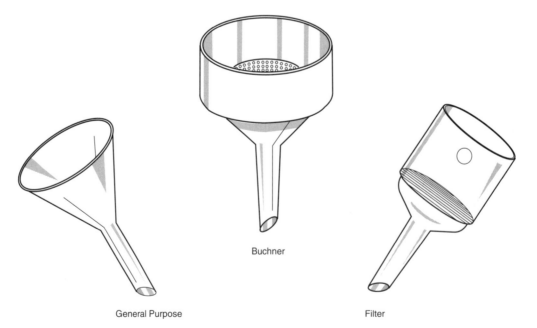

General Purpose   Buchner   Filter

Figure 1-25   Varieties of funnels

for bacteriological testing. The Durham tube is inverted to trap fermentation gases from bacteriological growth. Several types of stoppers are available for test tubes. Cotton plugs or stoppers that could react with tube contents, i.e., metal stoppers with acids, should not be used.

Thermometers. Thermometers are calibrated for total or partial immersion. Total immersion thermometers must be completely immersed in water to yield the correct temperature. Partial immersion thermometers must be immersed only to the depth of the etched circle that appears around the stem just below the scale level. An all-metal dial thermometer may also be used but must be calibrated. A thermometer in an oven can be placed in a flask filled with sand to protect from breakage and from temperature changes when the door is opened.

An electronic thermometer uses a thermistor and a digital readout. This type of thermometer must also be calibrated. Follow the manufacturer's instructions for use and calibration. For best results, check the accuracy of the thermometer in routine use against a thermometer certified by the National Institute of Standards and Technology.

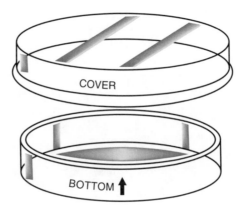

Figure 1-26   Petri dish

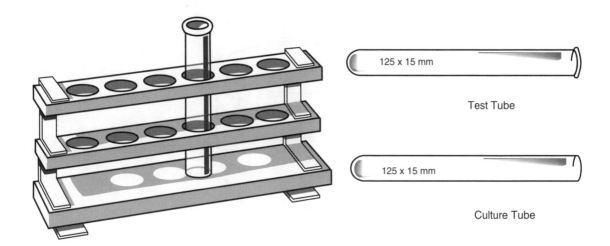

Figure 1-27   Test tubes and storage rack

Aspirator. The T-shaped attachment (shown in Figure 1-28) to the sink faucet is a relatively inexpensive vacuum device used instead of a vacuum pump to filter samples.

Desiccators. A desiccator is a container of heavy glass with a removable top and a false bottom above the real base (see Figure 1-29). The space below the false bottom is filled with a drying agent to keep the air dry while hot dishes cool to room temperature. A 12-in. diameter desiccator provides enough space for most water plant demands.

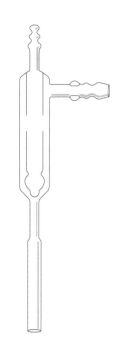

Figure 1-28   Aspirator

Figure 1-29   Desiccator

Weighing and Measuring Equipment

Many tools in the laboratory are used for specialized measuring or weighing. They are designed to give extremely accurate measurements and weights of chemicals and liquid. Such accuracy is essential in laboratory analysis.

Balance. An ordinary electronic analytical balance (shown in Figure 1-30) weighs substances as heavy as 200 g with an accuracy of 0.0001 g or one-tenth of a milligram. Such careful measurement is essential for preparing standard solutions. Follow the manufacturer's directions carefully to ensure correct weight and balance sensitivity.

Periodically calibrate all balances using a set of S or S-1 weights with the range of checks that encompass the range of weights being measured. Routine checks should include a sensitivity check to confirm that the balance can determine the difference between 150.0 g and 150.1 g or a similar difference near the range of weights normally measured by the balance. If checks confirm a problem, balance maintenance is required. Place balances on heavy tables made to resist vibration (see Figure 1-31).

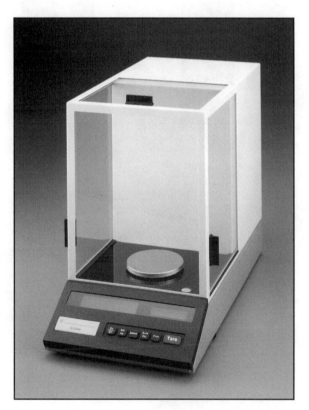

Courtesy of Denver Instrument

Figure 1-30   Electronic balance

The triple beam balance is also commonly used in laboratories to weigh larger quantities to less exacting standards than the electronic balance (see Figure 1-32). When the balance is idle, remove the balance beam and pans from their knife-edge supports with the knob. Protect the balance from air currents that may cause the pointer to weave, and zero-adjust it regularly. A good housekeeping practice is to clean the balance pans with a camel-hair brush, before and after using any balance.

Add weights one at a time, beginning with the heaviest and continuing down to the next consecutive lighter weight. When adding weights of 1 g or more, lift the beam from its knife edges. When adding weights smaller than 1 g, use the pan rests. Always return the weights to their proper places in the weight box.

Figure 1-31   Balance table

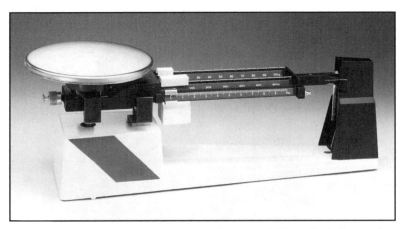

Courtesy of Ohaus Scale Corporation

Figure 1-32   Beam balance

A set of glass or plastic pans is desirable for weighing most solid chemicals needed to prepare standard solutions. These pans save the metal pans of the balance from being attacked by corrosive chemicals. Place one pan on the metal pan of the balance and adjust the weighing mechanism to exact zero. Transfer the required analytical weights either automatically (as with the electronic balance) or by hand (for the beam balance). Add the chemical with a spatula to the glass pan until it contains the correct quantity.

General Purpose Laboratory Equipment

The equipment discussed in this section is used for steps in a variety of laboratory procedures. For example, the oven is used for drying dissolved solids samples and for preparing glassware for specific analyses.

Autoclaves. Autoclaves, illustrated in Figure 1-33, are used to sterilize laboratory equipment, bacteriological media, and contaminated waste with pressurized steam. Use a maximum registering thermometer, autoclave tape, or spore strips with each autoclaving cycle to ensure sterilization temperatures have been reached.

Microscopes. Microscopes magnify extremely small objects so they can be seen (and often counted). Microscopic examination of water yields information on the presence and amounts of plant and animal life, such as algae, diatoms, protozoa, and crustacea.

Because microscopes are available in a variety of basic types, optional equipment, and prices, the purchase of a microscope requires research into the laboratory's needs and the best supplier to meet those needs. The manufacturer's representative should maintain the microscopes and instruct laboratory staff in proper use. The manufacturer's instructions should be followed carefully. Figure 1-34 shows a type of microscope typically found in water plant laboratories.

Muffle furnace. This high-temperature oven, shown in Figure 1-35, ignites and burns volatile solids. It is usually operated at temperatures near 600 °C.

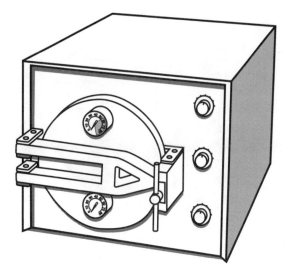

Figure 1-33   Autoclave

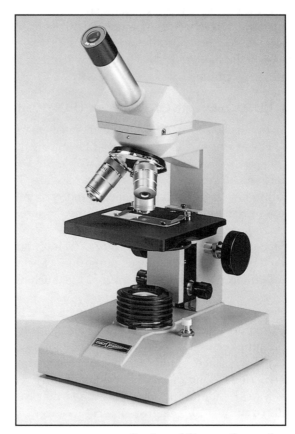

Courtesy of Olympus Corporation

Figure 1-34   Microscope

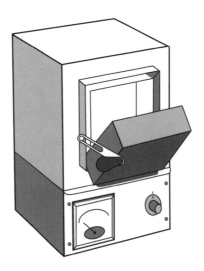

Figure 1-35   Muffle furnace

Oven. Ovens are used in the laboratory to dry, burn, or sterilize substances. An example of an oven typically found in laboratories is shown in Figure 1-36.

Vacuum pump. The pump in Figure 1-37 creates a vacuum for filtering laboratory samples. The membrane filter apparatus assembly (used with either a vacuum pump or aspirator) is shown in Figure 1-38.

Specific Purpose Laboratory Equipment

The laboratory equipment described here is used for specific procedures, for example, to measure conductivity, pH, for jar testing, and color comparisons, and nearly all

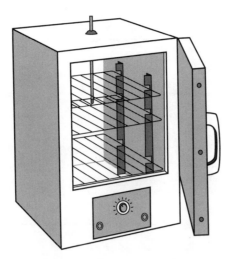

Figure 1-36   Oven

Courtesy of Gast Manufacturing Company

Figure 1-37   Vacuum pump

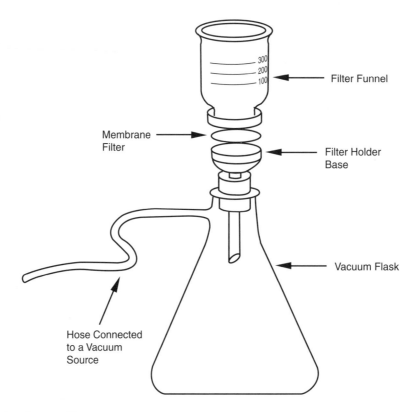

Figure 1-38   Membrane filter apparatus

need calibration to ensure the accuracy of their measurements.

Conductivity meters. Conductivity meters, such as the one shown in Figure 1-39, measure the ability of a solution to conduct an electric current. Generally, the lower the conductivity, the purer the water.

Jar testing equipment. Essential to conventional coagulation/filtration water treatment plants, this equipment features six jars, either square or round, and variable speed paddles. Optional background light is particularly helpful when viewing small floc. See Figure 1-40.

pH meter. This essential piece of laboratory equipment is a voltmeter that indicates acidity (less than 7.0) or alkalinity (more than 7.0) by measuring the hydrogen–ion activity in a solution. Figure 1-41 shows a pH meter setup with the solution moving in the flask by means of a stir bar and magnetic stirrer. Follow the manufacturer's instructions for calibrating the pH meter with standard buffers.

Turbidimeter. A turbidimeter (left in Figure 1-42) consists of a nephelometer with a light source to illuminate the sample and one or more photoelectric detectors with a readout device to indicate the intensity of light scattered at right angles to the path of the incident light.

Instrument sensitivity detects turbidity differences of 0.02 ntu or less in waters with a turbidity of less than 1 ntu. The range of the instrument should be from 0 to 10 units. Several ranges may be necessary to achieve this coverage.

Color comparisons. Several determinations in this manual specify the use of Nessler tubes for making color comparisons. For best results, place the sample in a tube identical to those of the color standards. The complete set of tubes should be matched; that is, they should be of the same size and have the same length of viewing path. Make the color comparison by looking into the tubes against a white surface to allow light to be reflected through the columns of liquid. Nessler tube sets are available in tall or short sizes. Usually tall is preferable. If color standards are to be retained for an extended time, cover the Nessler tube with transparent plastic film secured by rubber bands to protect against evaporation and contamination. This

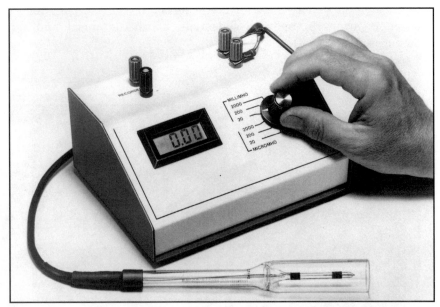

Courtesy of YSI, Incorporated

Figure 1-39   Conductivity meter (or bridge)

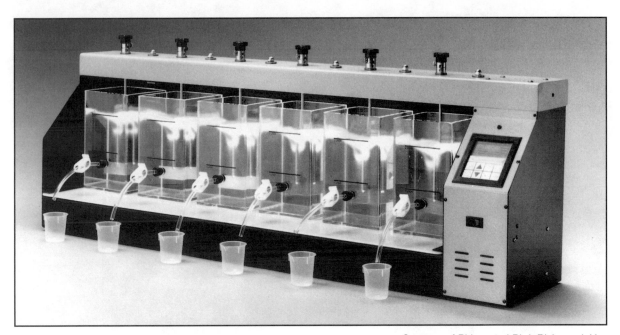

Courtesy of Phipps and Bird, Richmond, Va.

Figure 1-40   Jar testing equipment

arrangement enables the colors to be seen through the plastic film.

In many instances, sample colors are developed directly in the Nessler tubes in which the final comparison is made. After adding the reagents, mix the contents of the Nessler tube thoroughly. Close the tubes with clean, rinsed rubber stoppers and invert them four to six times for an adequate mix. Another way of mixing is with a plunger made from a glass rod, one end of which is flattened in a gas flame.

Comparators and test kits. Like standard solutions, permanent standards for color, turbidity, and many chemical substances are available commercially. The colorimetric kits come in two general forms: the disk type, which contains a color wheel (Figure 1-43), and the slide type, which contains liquid standards in glass ampules (Figure 1-44).

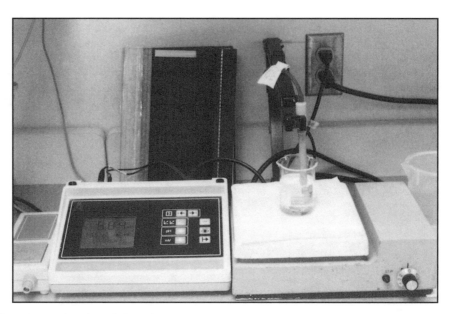

Figure 1-41   pH meter setup in a laboratory

Figure 1-42   Turbidimeter (l) and spectrophotometer (r)

The disk comparator consists of a plastic box with an eyepiece in front and a frosted glass in the rear. Behind the eyepiece is a place for attaching the rotating color disk. Between the disk and the frosted glass is a divided compartment for two cells that accommodate both the untreated water sample and the reagent-treated sample. Position the untreated water sample on the same side and in line with the rotating permanent colors. Reserve the companion side of the compartment for the sample in which the color has been chemically developed. Estimate the concentration visually by peering through the eyepiece and matching the developed color with the permanent colors on the disk. Remove the color disk from the kit and replace it with

Courtesy of Hach Company

Figure 1-43   Color comparison kit with color wheels for several analyses

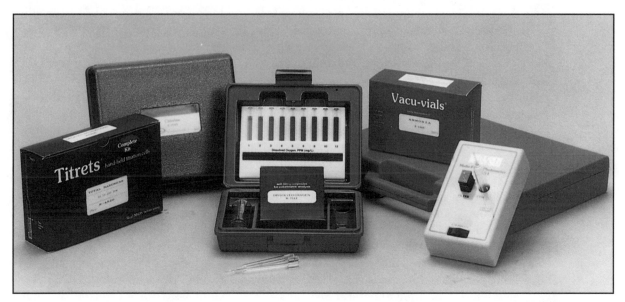

Courtesy of CHEMetrics, Inc.

Figure 1-44   Color comparison kit, featuring ampoule comparators, for dissolved oxygen

a disk for another determination. Thus, one comparator kit can serve for several determinations.

Check all these kits for reliability against a standard immediately after purchase and periodically thereafter. The standard solutions described in this manual are suitable for checking the calibration of such permanent standards. Most kits are designed for use with the manufacturer's own reagents, so new batches of reagents must be purchased as the need arises. If results are questionable, always recheck any newly arrived batch of reagents.

Always be alert for any deterioration of the purchased reagents after prolonged standing. If the reagents fail to yield reasonable results with the standard solutions, replace them immediately. When a test is performed at infrequent intervals, buy the needed solutions and reagents in small amounts, and date the bottles on delivery so that inaccurate results based on use of out-of-date reagents may be avoided. Generally, stable reagents are marketed in solution form; unstable reagents are put up in the form of a standard-size pill or a powder that can be dispensed with a small measuring spoon. The pill or powder may be added in the dry form or dissolved in a definite volume of water to prepare enough solution for one test.

Test kits give rapid, acceptable, and consistent results for people with minimal training. Their easy portability makes them useful for checking operations in the field. Many test kits are based on a simplified version of a test from *Standard Methods*. The reagents come ready for use and usually bear a code number or trade name, both of which help in ordering replacements. Among the advantages of a test kit are freedom from having to make up solutions; no need to prepare standards for every occasion; and an apparatus assembly designed specifically for the determination. However, the accuracy of tests run by kits may not equal that attainable in a good laboratory. Precision (the ability to reproduce the same result time after time) may approach the results obtained in the laboratory. Test kits should be used to monitor source and treated water only after trials demonstrate that the results on those particular waters equal or closely approximate values obtained by recognized and accepted standard methods.

Test kits and reagent solutions can be purchased for the following determinations: aluminum, ammonia nitrogen, residual chlorine, chlorine dioxide, color, copper, fluoride, iron, manganese, pH, phosphate and polyphosphate, silica, turbidity, alkalinity, calcium, calcium carbonate stability test (available under the name of Enslow stability indicator), carbon dioxide, chloride, hardness, and dissolved oxygen. The last seven determinations listed involve titrations rather than colorimetric estimations.

Colorimeters. Colorimeters measure a specific constituent in a sample by measuring the intensity of color after reagents have been mixed with the sample. They operate on a specific color filter that the light beam passes through. Colorimeters that are expandable with other filters to measure more than one constituent are also available. See Figure 1-45.

Photometers. These instruments rely on a photoelectric estimation of the intensity of light transmitted through or absorbed by the sample. Simpler photometers come equipped with a series of color filters and calibration or scale cards for a number of determinations. Use the manufacturer's reagents and be on guard against questionable reagents.

Spectrophotometers. This equipment uses a prism or grating to control the light wavelengths used for specific analysis rather than a color filter as do colorimeters and photometers. See Figure 1-46. Light waves are directed through a cuvette, an optical-quality, glass or quartz cell, shown in Figure 1-47. Spectrophotometers measure the amount of electromagnetic radiation absorbed by a sample, as a function of wavelength.

Other types of comparator kits and photometric and spectrophotometric instruments are available commercially. A description of all models is beyond the scope of this manual; AWWA does not make recommendations regarding their suitability for a particular laboratory.

Other Laboratory Equipment

Stands. Stands support apparatus such as burets and funnels. Clamps attached to the stands allow a variable height.

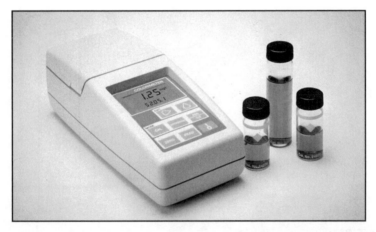

Courtesy of Hach Company

Figure 1-45   Colorimeter

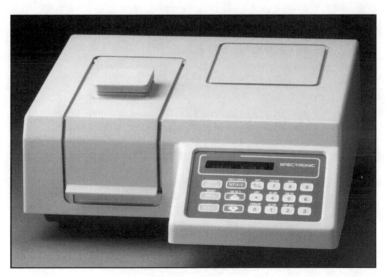

Courtesy of Milton Roy Company, Rochester, N.Y.

Figure 1-46   Spectrophotometer

Spatulas. Stainless steel and plastic spatulas are needed to transfer solid chemicals.

Heating procedures. Ordinary heating procedures are performed with Bunsen-type burners (shown in Figure 1-48), but many laboratories rely on an electric hot plate, which eliminates the ring stand necessary with a gas burner. In general, gas heat can be regulated from a low to a high temperature, whereas the cheaper electric heating units cannot be adjusted so finely.

Tongs and forceps. Tongs and forceps are useful for handling heated beakers and Erlenmeyer flasks. A wash–water bottle for rinsing beakers and Erlenmeyer flasks during quantitative transfers is also a necessity.

Marking pens. Laboratory marking pens are valuable for temporarily identifying sample bottles and the glassware in which tests are carried out.

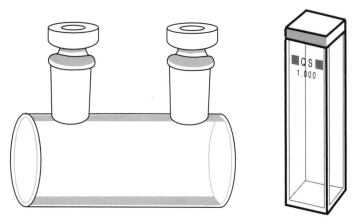

Figure 1-47   Cuvettes

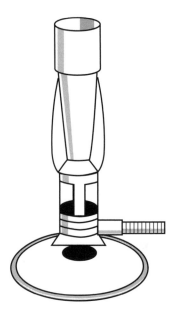

Figure 1-48   Gas burner (Bunsen or Fisher burner)

Reagent-grade water. Water is typically prepared by distillation, deionization, or reverse osmosis, followed by polishing with a mixed bed deionizer, and passage through a 0.2-μm pore membrane filter. See Figures 1-49 and 1-50.

RECORD KEEPING

Record reports and results of testing carefully to permit easy reference and to compare water conditions over time. It may be possible to save time in routine analyses by consulting previous records. Standard solution proportions, conversion factors, and other frequently used data should be available in a conveniently indexed form. Often a bound notebook is used to keep these records.

Regulatory agencies generally require water utilities to submit periodic reports on forms of the agencies' own design. Make copies of those forms for the water plant's files. Regulatory requirements frequently require that laboratory records be maintained for a specified length of time. Where laboratory records are considered legal records, they may be subject to the Freedom of Information Act regulations. For situations in which the water supply might be suspected of inferior quality, accurate record keeping will provide essential information for the water utility.

Most water utilities develop customized forms to record laboratory data. These may cover specific periods of time—day, week, or month. Computerized forms may also be provided in the laboratory. Keeping accurate and complete records allows the

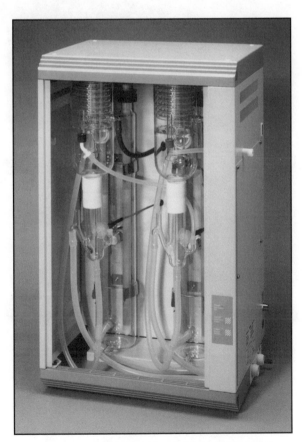

Courtesy of Barnstead/Thermolyne Corp.

Figure 1-49   Water still and deionizer cartridge

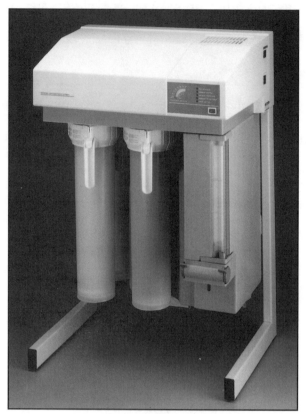

Courtesy of Barnstead/Thermolyne Corp.

Figure 1-50   Reverse osmosis system with cartridge polisher

water utility to assure customers that their water is being treated in the most efficient and effective manner possible. Records provide a basis from which to recognize trends and either resolve negative trends before they become a problem or refine positive trends to ensure the best possible quality drinking water for the community.

Chapter **2**

Alkalinity

PURPOSE OF TEST

Alkalinity is a measure of a water's ability to neutralize acids. In most drinking water, alkalinity is the result of bicarbonates (HCO_3), carbonates (CO_3^{2-}), and hydroxides (OH^-) of the metals calcium, magnesium, and sodium. This is shown by the relationship:

$$Total\ Alkalinity = (HCO_3^-) + 2(CO_3^{2-}) + (OH^-) - (H^+)$$

Alkalinity is expressed as total alkalinity, mg/L as calcium carbonate ($CaCO_3$). This alkalinity test provides results for use in interpreting and controlling water treatment processes. Many chemicals used in treating water can change its alkalinity; the most pronounced changes are caused by coagulants and softening chemicals. The results of this test are used to calculate chemical dosages needed in the coagulation and softening processes. Also, total alkalinity must be known for the Calcium Carbonate Saturation Test (see chapter 5) and for estimating carbonate hardness (see chapter 14).

LIST OF SIMPLIFIED METHODS

Refer to *Standard Methods for the Examination of Water and Wastewater*, Section 2320 B. Alkalinity, Titration Method. For more information on alkalinity analyses, see appendix A for US Environmental Protection Agency and American Society for Testing and Materials methods.

SIMPLIFIED PROCEDURE

Alkalinity Titration Method

Warnings/Cautions.

This method is suitable for titrating waters that contain hydroxide, carbonate, or bicarbonate alkalinity. Water should be free of color or turbidity that might obscure the indicator response. When water fails to satisfy any of these conditions, follow the procedures in *Standard Methods*.

Apparatus.

- a 25-mL buret and support

- a 100-mL graduated cylinder or volumetric pipet to measure the sample

- two or more 150-mL flasks or porcelain crucibles

- two or more stirring rods

- three dropping pipets or medicine droppers of 0.5-1 mL capacity for dispensing sodium thiosulfate ($Na_2S_2O_3 \cdot 5H_2O$) and indicator solutions

Reagents.

The following solutions are available commercially.

Sodium thiosulfate solution, $0.1N$ $Na_2S_2O_3 \cdot 5H_2O$ (not required if the water contains no residual chlorine). Dissolve 2.5 g sodium thiosulfate and dilute to 100 mL with distilled water.

Phenolphthalein solution, alcoholic, pH 8.3 indicator. Dissolve 0.5 g phenolphthalein disodium salt powder in 50 mL 95 percent ethyl or isopropyl alcohol and dilute to 100 mL with distilled water.

Bromcresol green indicator solution, pH 4.5 indicator. Dissolve 100 mg bromcresol green, sodium salt, in 100 mL distilled water.

Sulfuric acid titrant, $0.02N$ H_2SO_4. This solution requires some skill to prepare, standardize, and adjust to exactly $0.02N$ (consult *Standard Methods*).

Procedure.

1. Fill the buret to zero with sulfuric acid titrant and make sure there is no air in the buret before measuring. Record the liquid level in the buret by reading at the bottom of the meniscus. Make sure the stopcock does not leak.

2. Measure the sampler volume (Table 2-1) for the indicated alkalinity ranges:

    Example: If the alkalinity falls within the range of 0–250 mg/L as $CaCO_3$, take a 100-mL sample.

    Place equal volumes of the sample into two 150-mL flasks (or porcelain crucibles), one of which will be used as a blank for color comparison. Pour 100-mL sample into each flask.

3. If necessary, remove the residual chlorine by adding 1 drop (0.05 mL) sodium thiosulfate solution to each flask (or crucible) and mix.

4. Add 2 drops of phenolphthalein indicator solution to one flask (or crucible) and mix. If the sample turns pink, carbonate or hydroxide is present; proceed with step 5. If the sample remains colorless, the water contains bicarbonate, skip steps 5 to 7 and go on to step 8.

5. If the sample turns pink, gradually add sulfuric acid titrant from the buret, shaking the flask (or stirring the contents of the porcelain crucible) constantly until the pink just disappears. Use the flask without the phenolphthalein indicator as the color comparison blank.

6. Read the new buret level at the bottom of the meniscus and calculate the volume of acid used by subtracting the present buret reading (step 1) from the initial reading.

7. Calculate the phenolphthalein alkalinity P in terms of mg/L as calcium carbonate by multiplying the result found in step 6 by the appropriate factor. See Table 2-2.

   Example: If the titration requires 3 mL of titrant to reach the phenolphthalein end point for a 100-mL sample, multiply 3 by a factor of 10 to get a phenolphthalein alkalinity of 30 mg/L as $CaCO_3$.

8. Add 2 drops (0.1 mL) of bromcresol green indicator solution to both flasks (or crucibles) that contain the water sample.

9. Again titrate with small volumes of sulfuric acid until the bluish color begins to change to a bluish green. Continue titrating until a greenish color appears. The color change takes place within the span of 2 to 4 drops of sulfuric acid. Yellow means titration has gone beyond the end point. Use the comparison flask to help identify the color change.

10. Again read the buret and calculate the total volume of acid used in both the phenolphthalein titration (step 5, if carried out) and the bromcresol green titration (step 9). Multiply by the factor given in step 7. The result is the total alkalinity T in terms of mg/L as calcium carbonate.

Table 2-1   Determining sample volume by alkalinity range

Sample Volume mL	Alkalinity Range mg/L as $CaCO_3$
100	0–250
50	251–500
25	501–1,000

Table 2-2   Factors to calculate phenolphthalein alkalinity as calcium carbonate

Sample Volume	Factor
100	10
50	20
25	40

Example: If the titration requires 9.3 mL of titrant to reach this end point in a 100-mL sample, multiply 9.3 by 10 to get a total alkalinity of 93 mg/L as calcium carbonate.

11. Use Table 2-3 to calculate alkalinity relationships, where P is phenolphthalein alkalinity and T is total alkalinity.

Table 2-4 may be used with Table 2-3 to indicate relative proportions of hydroxide, carbonate, and bicarbonate alkalinity. For example, if total alkalinity is 76 mg/L and phenolphthalein alkalinity is 2 mg/L, then hydroxide will be 0, carbonate will be 4, and bicarbonate alkalinity will be 72 mg/L.

Table 2-3   Calculating alkalinity from titration results[*]

Titration	Hydroxide	Carbonate	Bicarbonate
$P = 0$	0	0	T
$P < 1/2T$	0	$2P$	$T-2P$
$P = 1/2T$	0	$2P$	0
$P > 1/2T$	$2P-T$	$2T-2P$	0
$P = T$	T	0	0

*P = phenolphthalein alkalinity (mg/L); T = total alkalinity (mg/L).

Table 2-4   Relative proportions of various alkalinities

Examples:[*]	1	2	3	4	5
If:					
Total alkalinity T	76	76	76	76	76
Phenolphthalein alkalinity P	0	2	38	42	76
Then:					
Hydroxide alkalinity	0	0	0	8	76
Carbonate alkalinity	0	4	76	68	0
Bicarbonate alkalinity	76	72	0	0	0

*P = phenolphthalein alkalinity (mg/L); T = total alkalinity (mg/L).
All values are in mg/L as $CaCO_3$.

Chapter **3**

Aluminum

PURPOSE OF TEST

Aluminum, the earth's most abundant metal, is present in natural waters at levels that cause no significant health or operational problems. Water treatment plants may use aluminum salts (alum) as coagulants. It is excess aluminum that remains in the treated water that is of concern. An incorrect dosage can increase the cost of operation and pose health risks. Accurate aluminum analysis is more important than ever and should be performed by a certified laboratory with approved methodology for the determination of aluminum in water. Colorimetric analysis (see reference section) may be performed by plant personnel to estimate aluminum concentration (greater than 0.05 mg/L) for correct coagulant dosage.

LIST OF SIMPLIFIED METHODS

The Eriochrome Cyanine R method provides the only means for aluminum analysis with simple instrumentation. This method has been adapted by commercial manufacturers for on-line analyzers. Reagents and standards can be purchased pre-packed for routine use. Colorimeters for direct concentration readout are available with kits and simplified instructions (see appendix A for resources). This is a simplified method. The US Environmental Protection Agency recommends the use of atomic absorption (AA), AAGF, AAPF, inductively coupled plasma (ICP), and ICP mass spectrometer techniques.

Refer to *Standard Methods for the Examination of Water and Wastewater*, Section 3500-Al B. Eriochrome Cyanine R Method.

SIMPLIFIED PROCEDURE _____

Eriochrome Cyanine R Method

Warnings/Cautions.

Negative errors are caused by fluoride and polyphosphates. The sample should be free of color and turbidity. Consult *Standard Methods* or follow the manufacturer's instructions for removal.

Some chemicals used in this procedure may be hazardous if mishandled or misused. Read all warnings on the reagent labels and follow laboratory safety procedures. Wash thoroughly if contact occurs.

Apparatus.

- spectrophotometer, for use at 535 nm, with a light path of 1 cm or longer

- colorimeter kit (may be used in lieu of spectrophotometer)

- glassware cleaned with warm 1:1 hydrochloric acid (HCl) and rinsed with aluminum-free distilled water

- measuring pipets

- volumetric pipets

- graduated cylinders

- Erlenmeyer flasks

- volumetric flasks

Reagents.

Aluminum standard solutions. Purchase a pre-made 1,000 mg/L stock solution. Pipet 1 mL of stock solution, using a volumetric pipet, into a 1,000-mL volumetric flask and dilute to the mark with deionized water. This aluminum working solution is 1,000 mg/L (1 mL = 1 mg Al). Prepare a set of at least three standard solutions using the table below. (Also see the Standard Calibration Curve section in chapter 1). The aluminum concentrations of these standards should cover the anticipated range of sample concentrations. The optimum aluminum range for this method lies between 20 and 300 mg/L. Samples may be diluted if aluminum concentration exceeds 300 mg/L. Standard aluminum solutions may also be available for purchase.

Because the remainder of the reagents are specific to each vendor's "kit," it is necessary to refer the analyst to the information provided by the vendor for completion of the analysis.

Procedure.

Refer to vendor-supplied information for the appropriate procedure for the "kit" being used.

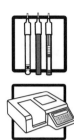

Chapter **4**

Ammonia–Nitrogen

PURPOSE OF TEST

Ammonia–nitrogen (NH_3–N) is an important parameter used to evaluate surface water quality. Small amounts of ammonia occur naturally, but a sudden increase in concentration may indicate sewage or industrial pollution. A rise in ammonia may be closely followed by an algae bloom with associated taste and odor problems. Over many years the water quality of a lake with high levels of ammonia will degrade, eventually becoming unfit as a water supply.

Ammonia is also used by some water treatment plants in conjunction with chlorine to form the disinfectant chloramine.

LIST OF SIMPLIFIED METHODS

The ammonia-selective electrode method uses a special ammonia ion-sensitive electrode and a pH meter with expanded millivolt scale to measure ammonia levels that range from 0.03 to 1400 mg/L ammonia–nitrogen (see *Standard Methods for the Examination of Water and Wastewater*, Section 4500–NH_3).

One common method formerly in widespread use was the Nessler method. It was a colorimetric analysis developed around the reaction of the Nessler reagent and ammonia, with a range from 0.02 to 2.0 mg/L ammonia–nitrogen. Its use was limited to purified drinking water, natural water, and highly purified wastewater with low color, turbidity, and an ammonia nitrogen concentration exceeding 0.02 mg/L. This method was dropped from the 19th edition of *Standard Methods* (because of the presence of mercury in the Nessler method), and its use by analytical laboratories is decreasing.

The salicylate method is also a colorimetric analysis but is not as widely used as the Nessler method (4500–NH_3 C. Titrimetric Method is an approved method under Clean Water Act [CWA] and not available for certification under Safe Drinking Water Act [SDWA]).

SIMPLIFIED PROCEDURES

Ammonia–Selective Electrode Method

Warnings/Cautions.

10N sodium hydroxide (NaOH) is used in this procedure. **NOTE: Sodium hydroxide is very caustic and should be prepared and handled with care.** All safety precautions outlined in the Material Safety Data Sheets (MSDS) should be followed. MSDS are provided by the chemical manufacturer or vendor. Interferences for this method include amines (that give a high reading) and mercury and silver (that are negative interferences). High concentrations of dissolved ions affect the measurement, but color and turbidity do not. Sample distillation is unnecessary for this procedure.

Apparatus.

* specific ion meter or pH meter with expanded millivolt scale capable of 0.1 mV resolution from –700 mV to +700 mV

* ammonia-selective electrode

* magnetic stirrer, thermally insulated, with a TFE-coated stir bar

Reagents.

Ammonia-free water. Ordinary distilled water may contain some ammonia nitrogen. To remove this ammonia, two procedures are available. In the first, add 0.1 mL concentrated sulfuric acid (H_2SO_4) per L of distilled water, then redistill. In the second, pass distilled water through a column that contains strongly acidic cation exchange resin.

Sodium hydroxide solution. Prepare 10N NaOH in an ice bath by dissolving 400 g sodium hydroxide in 800 mL single–distilled water. Add sodium hydroxide slowly, stirring constantly. Work under a fume hood and surround the beaker with running water or ice water. Cool and dilute to 1 L with distilled water. This solution is available commercially.

Stock ammonium chloride solution. Dissolve 3.819 g anhydrous ammonium chloride (NH_4Cl) dried at 100°C, in 1 L ammonia-free water. 1 mL stock solution = 1 mg N = 1.22 mg ammonia (NH_3).

Working standards. Prepare a series of standards for the following concentrations: 1,000, 100, 10, 1, and 0.1 mg/L ammonia–nitrogen. Make decimal dilutions of the stock ammonium chloride solution with ammonia-free water. For more information on preparing a standard curve, see the Standard Calibration Curves section in chapter 1.

Procedure.

1. Test the electrode sensor performance (following the manufacturer's instructions) before conducting tests.

2. After preparing the working standards, calibrate the meter. Place 100 mL of each standard solution into a 150-mL beaker. Beginning with the standard of lowest concentration, place the electrode into the solution and

mix using the magnetic stirrer. Use caution while mixing to ensure that air bubbles are not pulled into the solution and trapped on the electrode membrane. Maintain a constant temperature and rate of stirring throughout the calibration and testing procedures.

3. Add 1 mL of sodium hydroxide solution to the standard solution being mixed. The pH of the standard solution should rise above pH 11. Do not add sodium hydroxide to the solution before immersing the electrode, because ammonia may be lost from the basic solution.

4. Keep the electrode in the solution until a stable millivolt reading is reached. Record this reading.

5. Repeat the above procedure with the remaining standard solutions, beginning with the lowest concentration and working up to the highest. For standards or unknown samples, wait at least 5 min before recording the millivolt reading.

6. Using the data collected from standards, prepare a calibration curve. On semilogarithmic graph paper plot ammonia concentration in mg/L ammonia–nitrogen on the log axis versus potential in millivolts on the linear axis. If the electrode is functioning properly, a tenfold change in ammonia concentration should produce a potential change of about 59 mV. If using a specific ion meter, refer to the manufacturer's instructions and proceed with the calibration as outlined in steps 1–4.

7. To determine ammonia concentrations in unknown samples, place 100-mL sample in a beaker and follow the same procedure outlined earlier. Record the millivolt reading and the volume of $10N$ sodium hydroxide added in excess of 1 mL. Read the concentration of ammonia–nitrogen from the calibration curve at the millivolt reading for that sample. If the millivolt value is not on the calibration curve, dilute the sample and repeat the analysis.

8. Calculations. mg/L NH_3–N = $A \cdot B$ [(101 + C)/101]

Where:

A	=	dilution factor
B	=	concentration of NH_3–N in mg/L from the calibration curve
C	=	volume (in mL) of $10N$ NaOH added in excess of 1 mL

9. Alternately, follow manufacturer's instructions for calibration and record ammonia concentrations directly from the meter in mg/L.

This page intentionally blank

Chapter **5**

Calcium Carbonate Saturation

PURPOSE OF TEST

Calcium carbonate ($CaCO_3$) saturation indices are used to evaluate the scale-forming or scale-dissolving tendencies of water. Assessing these tendencies is useful in corrosion control programs and in preventing calcium carbonate scaling in piping or domestic water heaters.

Several water quality characteristics may need to be measured to calculate the calcium carbonate saturation indices. Minimum requirements include total alkalinity, pH, carbon dioxide, and temperature.

LIST OF SIMPLIFIED METHODS

Refer to *Standard Methods for the Examination of Water and Wastewater*, Section 2330 B. Indices Indicating Tendency of a Water to Precipitate $CaCO_3$ or Dissolve $CaCO_3$.

SIMPLIFIED PROCEDURE

Calcium Carbonate Stability Method

Warnings/Cautions.

The sample should be collected with as little splashing or aeration as possible to prevent loss of carbon dioxide.

In general, calcium carbonate tends to precipitate from oversaturated waters and to dissolve from undersaturated waters. In saturated waters, it tends to do neither. Exceptions may occur when polyphosphates or certain naturally occurring organic compounds or magnesium are present. Do not consider saturation indices as absolutes. Instead, view them as guides to the behavior of calcium carbonate in aqueous systems and supplement them with experimentally derived information where possible.

Apparatus.

- 300-mL glass-stoppered biochemical oxygen demand (BOD) bottles
- a 100-mL pipet
- filter funnel
- filter paper (Whatman #50 or equivalent)
- apparatus for determining alkalinity, carbon dioxide, pH, and temperature

Reagents.

Calcium carbonate ($CaCO_3$), precipitated powder, reagent grade
Reagents for determining alkalinity (see chapter 2)

Procedure.

1. Determine total alkalinity of the water sample (see the Alkalinity Test Procedure in chapter 2).

2. Collect a second sample in a BOD bottle without splashing or agitating. Completely fill the bottle and stopper (see the carbon dioxide test procedure in chapter 6 for proper technique).

3. Add approximately 0.3 to 0.4 g calcium carbonate powder to the bottle.

4. Carefully replace the stopper so no air bubbles are left.

5. Mix the powder into the sample by shaking the bottle every 10 min for at least 3 hr.

6. Allow the sample to settle overnight. During the first part of the settling, tap the bottle gently and twist the stopper so any powder that adheres to the walls and to the stopper can be loosened and allowed to settle to the bottom.

7. With a 100-mL pipet, carefully remove two portions (200 mL) of the supernatant (the clear layer above the settled material).

8. Filter the 200-mL portion of supernatant through filter paper with a filter funnel. Discard the first 25 mL of the filtrate (the sample that has passed through the filter paper) and save the rest.

9. Determine the total alkalinity of the filtrate (see the alkalinity test procedure in chapter 2). Make sure that the calcium carbonate powder is completely removed from the filtrate to avoid any error in the total alkalinity test.

Practical applications of the test are as follows:

- Water is unsaturated with respect to calcium carbonate and may be corrosive if the second total alkalinity result (step 9) is greater than the first (step 1).

- The water is oversaturated with calcium carbonate and may deposit a protective coating in the mains if the first result (step 1) exceeds the second (step 9).

- The water is stable and in equilibrium with calcium carbonate if the first (step 1) and the second (step 9) results are similar. These waters should be noncorrosive if there is already a calcium carbonate coating in the mains.

Changes in the sample's total alkalinity can occur with the following amounts of chemicals:

- Each 1 mg/L dose of pure lime (CaO) or 1.1 mg/L dose of 90 percent commercial quicklime increases total alkalinity by 1.79 mg/L or reduces carbon dioxide by 1.57 mg/L.

- Each 1 mg/L dose of pure hydrated lime [Ca(OH)$_2$] or 1.41 mg/L dose of 93 percent commercial hydrated lime increases the total alkalinity by 1.35 mg/L or reduces the carbon dioxide by 1.19 mg/L.

- Each 1 mg/L dose of pure soda ash (Na$_2$CO$_3$) or 1.02 mg/L dose of 98 percent commercial soda ash increases the total alkalinity by 0.944 mg/L or reduces the carbon dioxide by 0.415 mg/L.

These quantities will not necessarily stabilize every water because stability is interrelated with other important factors. For a more extended discussion of this test and its applications, see the bibliography in the reference provided for *Standard Methods*.

Scaling of pipes and equipment by water saturated with calcium carbonate may be minimized by adding phosphate salts, stabilizing the softened water through recarbonation, adding alum, or contact with limestone.

This page intentionally blank

Chapter **6**

Carbon Dioxide (Free)

PURPOSE OF TEST

Because of the corrosive properties of dissolved carbon dioxide gas, it is desirable to eliminate it from drinking water. Carbon dioxide can be removed either by aeration or through chemical conversion to the less aggressive bicarbonate or carbonate states by adding alkaline compounds.

A knowledge of carbon dioxide content is also important in softening treatment. First, carbon dioxide consumes additional lime and soda ash as it is neutralized. Second, carbon dioxide may be applied to softened water just before filtration to dissolve any unsettled calcium carbonate that might otherwise deposit later on filter sand or in distribution mains. As a rule, groundwaters contain more dissolved carbon dioxide gas than do surface waters. In most drinking water supplies, carbon dioxide represents the important acid factor, making titration for carbon dioxide substantially equal to acidity titration.

LIST OF SIMPLIFIED METHODS

Refer to *Standard Methods for the Examination of Water and Wastewater*, Section 4500–CO_2 C. Titrimetric Method for Free Carbon Dioxide. The titration method presented is the only commonly used method for determining carbon dioxide in a sample with readily available laboratory equipment. The sample is titrated with a solution that contains sodium carbonate in a reacting concentration equivalent to that of carbon dioxide gas.

Kits are available for this test. Determine whether the kit contains only the reagents or the reagents plus equipment to perform the test, then choose according to needs (see appendix A for a list of manufacturers and suppliers).

SIMPLIFIED PROCEDURE

Sodium Carbonate Titration Method

Warnings/Cautions.

Because carbon dioxide is a gas, be careful to avoid any vigorous agitation of the sample to prevent dissolved carbon dioxide from escaping. Carbon dioxide is present in air and can dissolve in a sample; therefore, avoid any undue exposure of the sample to air before testing. Collect each sample with as little splashing and aeration as possible to prevent the loss of dissolved carbon dioxide gas.

Do not collect the samples listed later in this chapter in Titration of Sample, until the procedure states to do so. Best results are achieved if each step in the test is performed on a freshly collected sample.

If the determination must be delayed for any reason, a sample should be collected in a 500-mL bottle and stoppered to prevent air from being trapped between the stopper and the water sample.

When the carbon dioxide concentration exceeds 10 mg/L, repeat the titration on a fresh sample to check the reliability of the first value.

Apparatus.

- a supply of one-hole rubber stoppers for inserting in the water tap

- copper, stainless steel, or other noncorrosive metal tubing of proper diameter for inserting in the rubber stopper

- rubber or plastic tubing to connect to the metal tube

- four or more 100-mL graduated cylinders or Nessler tubes

- a 0.5- or 1-mL dropping pipet or medicine dropper for dispensing the phenolphthalein indicator solution

- a stirring rod long enough to reach the bottom of the 100-mL graduated cylinders or Nessler tubes

- a 25-mL buret and support to measure the sodium carbonate solution

- a 100-mL glass-stoppered bottle to store the phenolphthalein solution

- weighing paper or a small evaporating dish to weigh sodium bicarbonate

- a 250-mL beaker to dissolve sodium bicarbonate in boiled distilled water

- a 250-mL or larger graduated cylinder to measure water for dissolving sodium bicarbonate

- a 1-L volumetric flask to prepare the sodium bicarbonate solution

- two glass funnels of 4-in. (100-mm) diameter to transfer sodium carbonate solution to a 1-L volumetric flask and a storage bottle

- a 1-L glass bottle to store sodium bicarbonate solution

- a rubber stopper to fit the 1-L storage bottle for sodium bicarbonate solution

- a 500-mL glass-stoppered bottle to collect samples to be transported to the laboratory for testing

- a length of rubber or plastic tubing to serve as a siphon from the bottom of the 500-mL glass-stoppered bottle to the bottom of a 100-mL graduated cylinder or Nessler tube

Reagents.

Boiled distilled water. Because most distilled water contains some carbon dioxide, place ordinary distilled water in a large unstoppered pyrex flask or bottle and boil for at least 15 min to expel the carbon dioxide. Then cover the top and neck of the vessel with an oversized inverted beaker and cool the water to room temperature in a bath of cold running water. Prepare the boiled distilled water immediately before preparing the standard sodium carbonate titrant.

Phenolphthalein solution, alcoholic, pH 8.3 indicator. Dissolve 0.5 g phenolphthalein disodium salt powder in 50 mL 95 percent ethyl or isopropyl alcohol, and dilute to 100 mL with distilled water. This solution is available commercially.

Sodium carbonate titrant, 0.0454N, Na_2CO_3. Discard this reagent 1 month after preparation. Prepare a fresh solution at the time of the next carbon dioxide test. In addition, solutions of the caustic titrant needed for this test can be purchased from carbon dioxide test kit suppliers. If the solution is sodium bicarbonate ($NaHCO_3$), the normality should be 0.0454 in order to apply the calculations specified in the titration sample section later in this chapter. If the solution is sodium hydroxide ($NaOH$), the normality may be different. In that case, use the calculation instructions provided by the kit manufacturer. The notes about discarding solution and the caustic properties of sodium carbonate also apply to sodium hydroxide.

1. On an analytical balance, carefully weigh 2.407 g dry sodium carbonate of primary standard-grade quality.

2. Carefully transfer to a 250-mL beaker and dissolve in 150-mL boiled distilled water.

3. Insert a clean funnel into the neck of a 1-L volumetric flask and carefully transfer the solution from the beaker to the flask.

4. Rinse the beaker with three 100-mL portions of boiled distilled water and add the rinses to the volumetric flask. Add boiled distilled water to the volumetric flask to the 1 L mark. Stopper the volumetric flask.

5. With a hand over the stopper, thoroughly mix the solution by slowly inverting the flask at least three times.

6. Use a clean funnel to transfer the solution from the flask to a 1-L glass bottle. Insert a clean rubber stopper into the mouth of the bottle. (A glass stopper may "freeze" to the mouth of the glass bottle.)

7. Label the bottle with the name and normality (0.0454 N) of the solution, the date of preparation, and initials. Also print "Caustic" on the bottle as a caution.

NOTE: **This solution is caustic and can burn skin on contact. If this occurs, immediately rinse the area of contact with cold tap water.**

Procedure.

Construct a special sampling line (see Figure 6-1) for collecting water that contains dissolved gases.

1. Select a one-hole rubber stopper that fits snugly into the inside of the water tap from which the sample is to be drawn.

2. Cut off a length of metal tubing equal to four times the length of the rubber stopper.

3. Wet the outside surface of the entering end of the metal tube with a lubricating film of glycerol or glycerine and carefully introduce the metal tube into the hole of the rubber stopper. Work the tube into the hole until the end of the tube is flush with the surface of the stopper.

4. Select rubber or plastic tubing that fits the metal tube and cut off enough length to reach from the tap to the bottom of the receiving vessel. The receiver will be a 100-mL graduated cylinder or Nessler tube, if the sample can be analyzed where it is collected. If the sample must be transported to the laboratory for analysis, the receiver will be a 500-mL bottle as shown in Figure 6-1.

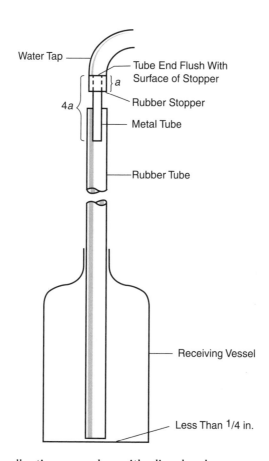

Figure 6-1   Apparatus for collecting samples with dissolved gases

5. Attach one end of the tubing to the projecting end of the metal tube. If necessary, again wet the outside surface of the metal tube with the lubricating film to ease the penetration of the rubber tubing.

6. Thoroughly clean the rubber stopper, metal tube, and delivery tubing of all traces of lubricating film.

Use the following procedure to collect all water samples that contain dissolved gases.

1. Insert the rubber stopper in the water tap. Make sure the entire sampling line is airtight and that no atmospheric oxygen can come into contact with the flowing water sample.

2. Flush the inside of the metal and rubber tubes with the water to be sampled.

3. Insert the open end of the delivery tubing to within $1/4$ in. (6 mm) of the inside bottom of the receiver—a 100-mL graduated cylinder or Nessler tube if the source of the sample is near the testing equipment. The receiver will be a 500–mL bottle if the sample is to be transported to the laboratory.

4. Let the water sample overflow the receiver several times the receiver capacity. Two or three min of overflow is usually sufficient.

5. Gently withdraw the rubber tubing from the receiver as the water continues to overflow.

6. If using a 500-mL bottle, immediately stopper the bottle, leaving no air space near the top. If there is an air space, remove the glass stopper, insert the delivery tube to within $1/4$ in. (6 mm) of the bottom of the bottle, collect more sample to overflowing, gently remove the delivery tube, and gently stopper the bottle. Repeat until there is no air space at the top of the sample in the stoppered bottle.

7. To minimize the escape of dissolved carbon dioxide gas from the bottled sample, keep the bottle at a temperature lower than that at which the water was collected. Titrate the sample as soon as possible after collecting the sample.

Titrating samples:

1. Collect the first sample as previously described. If the receiver is a 100-mL graduated cylinder or Nessler tube, flick the cylinder or Nessler tube to throw off the excess sample above the 100 mL mark.

   If the receiver is a 500-mL glass-stoppered bottle, use a siphon tube to transfer the sample from the bottom of the bottle to a 100-mL graduated cylinder or Nessler tube. To transfer the sample, place the tube down to the bottom of the bottle, begin the siphon flow, place the end of the siphon tube to within $1/4$ in. (6 mm) of the bottom of the 100-mL receiver. Allow the sample to overflow, then gently remove the siphon tube from the bottle and restopper. Flick the receiver to throw off any excess sample above the 100 mL mark.

2. Set the sample aside for later use as a color comparison blank in the titration.

3. Fill the buret with sodium carbonate titrant. Record the liquid level in the buret by reading at the bottom of the meniscus.

4. Collect a second sample as described in step 1.

5. Add 10 drops of phenolphthalein indicator solution to the second sample, or add powder from a purchased packet of this reagent, and stir gently with a long stirring rod. If the sample turns pink, no carbon dioxide is present. Record 0.0 mg/L CO_2 as the final test result and omit the next steps. If there is no color, wipe the stirring rod dry with a paper towel (do this after each use) and continue with the next step.

6. If the sample remains colorless, add sodium carbonate titrant to the cylinder or Nessler tube. Stir gently with a long stirring rod until a definite pink color persists for 30 sec. Look down the length of the cylinder or Nessler tube for evidence of the color change. Use the first sample with no phenolphthalein indicator solution as a color comparison blank.

7. Read the new buret level at the bottom of the meniscus and calculate the volume of sodium carbonate used by subtracting the initial buret reading (step 3) from the present reading.

8. Collect a third sample. Immediately add the full amount of sodium carbonate titrant found in step 7. Then add 10 drops of phenolphthalein indicator solution and stir gently with the long stirring rod. If the sample remains colorless, continue to add sodium carbonate until a definite pink color persists for 30 sec. Accept the new result as the reliable titration.

9. Calculate the mg/L free carbon dioxide concentration by multiplying the number of milliliters of sodium carbonate used by 10. Record the mg/L carbon dioxide test result.

NOTE: If you substitute a purchased titrant for the test, use this direction for the calculation only if the normality of the purchased sodium carbonate or sodium hydroxide is 0.0454 and you have used a 100-mL sample. If either or both conditions are different from the purchased titrant, use the supplier's instructions to calculate the test result.

10. Carefully clean all equipment used to collect the sample and perform the test by rinsing each item with tap water, brushing with soapy water, rinsing three times with tap water, then rinsing three times with distilled water. Drain to dry and store in a dust-free area.

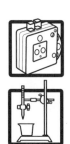

Chapter **7**

Chlorine Residual (General)

PURPOSE OF TEST

Chlorine is used primarily as a disinfectant to destroy disease-producing microorganisms in the water supply. In addition, it can improve the quality of finished water by reacting with ammonia, iron, manganese, sulfide, and some taste- and odor-producing substances.

Chlorine can also react with natural precursors producing undesirable disinfection by-products that may have adverse health effects. Excess chlorine also can cause taste and odor problems. Combined chlorine, formed when chlorine reacts with ammonia or amine compounds, can adversely affect aquatic life and persons with certain medical conditions.

Two types of chlorine residual are produced in water from the chlorination process: free available residual and combined available residual. *Free available residual* occurs when the water is thoroughly chlorinated. It exists in three forms: molecular chlorine (Cl_2), hypochlorous acid (HOCl), and hypochlorite ion (OCl^-). Molecular chlorine exists in the pH range of 1 to 4; hypochlorous acid in the pH range of 1 to 9 (it is the predominant form in the pH range of 2 to 7); hypochlorous acid and hypochlorite ion coexist in equal proportions at pH 7.4. Hypochlorite ion is the predominant form above pH 9.5. *Combined available residual* forms when chlorine reacts with ammonia that either occurs naturally or is added in treatment. It exists in three forms: monochloramine (NH_2Cl), dichloramine ($NHCl_2$), and trichloramine (NCl_3). Combined chlorine is a less effective disinfectant than free chlorine.

Chlorine residual analytical procedures measure free and total residuals, the sum of which is the total chlorine residual. Therefore, total chlorine residual is always equal to or greater than free chlorine residual.

LIST OF SIMPLIFIED METHODS

- field method using commercial comparator kit

- amperometric titration method (see appendix A for a list of manufacturers and suppliers)

- titrimetric method

- colorimetric method

For other methods, refer to *Standard Methods for the Examination of Water and Wastewater,* Section 4500–Cl:

- iodometric method

- low-level amperometric titration method

- syringaldazine (FACTS) method

- iodometric electrode method

SIMPLIFIED PROCEDURES

Chlorine (Residual) Field Method Using Commercial Comparator Kit

This simple test is designed for field or laboratory determination of free and total chlorine in water using N,N-diethyl-p-phenylenediamine (DPD) as the color indicator. DPD produces a pink color to a degree proportional to the chlorine content of the water. The color of the water is then compared to the standard color scale in the comparator to determine the chlorine content of the water.

In this procedure, a sample is added to two identical viewing tubes. DPD is added to one tube to produce a color change. The tube without the indicator serves as a control tube to ensure an accurate color match by nullifying color or turbidity in the sample. Match the color of the sample containing indicator to the standard color scale (which overlays the control tube) and read the chlorine content directly from the comparator.

Warnings/Cautions.

The buffer pillow lowers the sample pH within the range of 6.2 to 6.5 for accurate results. A lower pH enables chloramine to appear as free chlorine. A higher pH causes dissolved oxygen to give a pink color identical to that produced by chlorine.

If the sample contains oxidized manganese, an inhibitor (included in the kit) must be used.

High temperature enables chloramine to appear as free chlorine and increases color fading. Complete measurements rapidly at high temperatures.

Oxidized manganese that is naturally occurring or added during water treatment as potassium permanganate ($KMnO_4$) reacts with DPD to give a pink color identical to that produced by chlorine. In such a case a correction must be made for this interference. (See *Standard Methods* for further discussion.)

Chlorine dioxide (ClO_2), if present, appears with free chlorine to the extent of one-fifth the total chlorine content. (See *Standard Methods* for further discussion.)

Monochloramine (NH_2Cl), if present in high concentration, interferes in the free chlorine determination after 1 min of developing time. Therefore, all readings must be made within the specified time interval.

The DPD color comparator kit must be calibrated initially when the kit is purchased and thereafter at least once every 3 months. Color standards are light sensitive and fade when exposed to sunlight or high temperatures. Color comparators should be stored in a dark, cool location. Storing test kits inside vehicles shortens the life expectancy of color standards. Similarly, the shelf life of powder pillows or tablets is adversely affected when exposed to higher temperatures or direct sunlight. If the contents of the powder pillows or tablets are discolored, they should be discarded.

Apparatus.

- a color comparator kit complete with standard color scale (disk or blocks of appropriate chlorine range), color viewing tubes, and caps (commercially available)

- a 20–200-μL pipet

- a 1-L volumetric flask

Reagents.

DPD indicator. Commercially available as DPD Free Chlorine Reagent and DPD Total Chlorine Reagent; or prepare according to *Standard Methods*, Section 4500–Cl F. DPD Ferrous Titrimetric Method.

Distilled water. Use reagent-grade, deionized distilled water, which should be chlorine free.

Potassium permanganate stock standard, $KMnO_4$. Commercially available or prepare as follows: weigh 0.8910 g desiccated, reagent-grade potassium permanganate with an analytical balance. Transfer to a 1-L volumetric flask and bring to volume with distilled water. This solution is equivalent to 1,000 mg/L of chlorine. Using volumetric glassware, dilute 10 mL of the 1,000 mg/L stock with distilled water to a final volume of 100 mL. (This solution must be made fresh for each calibration.) This solution is equivalent to 100 mg/L chlorine.

Free chlorine determination procedure.

1. Rinse two viewing tubes with distilled water.

2. Fill a control viewing tube to the graduation mark with water to be tested, and place it in the opening of the color comparator behind the standard color scale.

3. Fill sample viewing tube to graduation mark with water to be tested.

4. Add entire contents of prepackaged free chlorine reagent to sample viewing tube. Cap and swirl to mix. (Occasionally, swirling alone does not produce good precision in data. To solve this problem, shake the sample vigorously for 15 sec, then swirl to remove air bubbles. The powder does not have to dissolve completely to obtain a correct reading.) Place the sample tube in the comparator beside the control viewing tube. The free chlorine must be read in 1 min.

5. Hold the comparator to a light source and compare the sample viewing tube to the comparator standard color scale. When a color match is achieved, record the free chlorine value in mg/L. The free chlorine must be read in 1 min.

Total chlorine determination procedure.

1. Rinse two viewing tubes with distilled water.

2. Fill the control viewing tube to the graduation mark with water to be tested and place in the opening of the color comparator behind the standard color scale.

3. Fill the sample viewing tube to the graduation mark with the water to be tested.

4. Add the entire contents of the prepackaged total chlorine reagent to the sample tube. Cap and swirl to mix. The powder does not have to dissolve completely to obtain correct readings. Place the sample tube in the comparator beside the control viewing tube. Let stand at least 3 min but not more than 6 min.

5. Hold the comparator to a light source and compare the sample viewing tube to the comparator standard color scale. When a color match is achieved, record the total chlorine value in mg/L.

Comparator calibration procedure.

Known amounts of a potassium permanganate solution are used as standards to verify the calibration of the comparator standard color scale used to determine chlorine concentration. For more information on preparing a standard curve, see the Standard Calibration Curves section in chapter 1.

Prepare standards as follows:

1. Fill a viewing tube to the graduation mark with distilled water.

2. Using a calibrated automatic microliter pipet, spike the tube with a known amount of potassium permanganate 100 mg/L chlorine equivalent solution.

3. Prepare at least six standards that span the entire chlorine range of the comparator standard color scale being calibrated. Determine the true concentrations of the standards by using Table 7-1.

NOTE: Because of different viewing tube volumes, the resulting concentration will vary for a given potassium permanganate spike volume. Process the standards as described previously and read the resulting color development against the comparator standard color scale. Compare the true value of the potassium permanganate standard with the observed value of the kit. Acceptable values are ± 10 percent of the true value. If concentrations are outside the acceptable range, repair or replace the kit, calculate a correction factor, or draw a correction curve to determine sample concentration.

Chlorine (Residual) Amperometric Method

This titration is designed primarily for laboratory rather than field use because it requires more skill and care than the colorimetric methods. Differentiation between

Table 7-1 Calibrating comparator standard using potassium permanganate

$KMnO_4$ Spike Volume	Viewing Tube Volume			$KMnO_4$ Spike Volume	Viewing Tube Volume		
µL	5 mL	10 mL	25 mL	µL	5 mL	10 mL	25 mL
5	0.10			90	1.80	0.90	
10	0.20	1.10		100	2.00	1.00	0.40
20	0.40	0.20		110	2.00	1.10	
25	0.50	0.25	0.10	125	2.50	1.25	0.50
30	0.60	0.30		140	2.80	1.40	
40	0.80	0.40		150	3.00	1.50	0.60
50	1.00	0.50	0.20	160	3.20	1.60	
60	1.20	0.60		170	3.40	1.70	
70	1.40	0.70		175	3.50	1.75	0.70
75	1.50	0.75	0.30	180	3.60	1.80	
80	1.60	0.80		190	3.80	1.90	
				200	4.00	2.00	0.80

free and combined chlorine is possible by pH adjustment and the presence or absence of potassium iodide (KI). Free chlorine can be determined at a pH between 6.5 and 7.5. Combined chlorine can be determined at a pH between 3.5 and 4.5 in the presence of the correct amount of potassium iodide.

Warnings/Cautions.

Phenylarsine oxide (C_6H_5AsO) titrant is a severe poison and a suspected carcinogen. Read all warnings regarding proper handling. Control of pH is important for correct results. Above pH 7.5, the reaction with free chlorine becomes sluggish. Below pH 6.5, some combined chlorine may react even in the absence of iodide. Below pH 3.5, oxidized manganese reacts with the titrant. Above pH 4.5, the titration for combined chlorine fails to reach completion.

High temperatures and prolonged titration time allow monochloramine to be titrated as free chlorine, which leads to an apparent increase in free chlorine results.

Chlorine dioxide and free halogens titrate as free chlorine, which leads to an apparent increase in free chlorine results. (See *Standard Methods* for further discussion.)

Excessive stirring of some commercial titrators can lower chlorine values by volatilization. Complete the analysis promptly.

Chlorine residuals higher than 2 mg/L are more accurately determined by using smaller sample volumes or by sample dilution.

NOTE: **Perform the analysis immediately after collecting the sample.**

Apparatus.

Amperometric titrator. A typical amperometric titrator consists of a two-electrode cell connected to a microammeter and an adjustable potentiometer. The cell unit includes a noble metal electrode, a reference electrode of silver–silver chloride in a saturated sodium chloride (NaCl) solution, and a salt bridge. An agitator and a buret are also required and are typically supplied with the titrator.

For best results, observe the following practices to prepare and operate the apparatus.

- Keep the noble metal electrode free of deposits. Occasional mechanical cleaning with a suitable abrasive is sufficient.

- Keep the salt bridge in good operating condition. If plugging or improper flow of salt solution occurs in the salt bridge, empty the old material from the cell and replace it with fresh salt.

- Keep an adequate supply of solid salt in the reference electrode at all times.

- Thoroughly clean the agitator and exposed electrode system to remove the chlorine-consuming contaminants by immersion for several minutes in water containing 1–2 mg/L free available residual chlorine. Add potassium iodide to the same water and immerse for another 5 min.

- Thoroughly rinse the sensitized electrodes and agitator with distilled water or the sample to be tested.

- If the chlorine concentration of the samples approximates 0.5 mg/L, condition the electrode system further by conducting two or more titrations at the 0.5 mg/L level until the titrations become reproducible.

- Satisfy the chlorine demand of all glassware to be used for sampling and titrating samples by subjecting the critical surfaces to a water that contains 10 mg/L or more residual chlorine for at least 3 hr. Rinse with distilled water to remove the residual chlorine traces.

Reagents.

Phenylarsine oxide titrant, C_6H_5AsO (0.00564N) (PAO). Purchase this reagent, or prepare as follows. Dissolve approximately 0.8 g phenylarsine oxide powder in 150 mL 0.3N sodium hydroxide solution. After settling, decant 110 mL into 800 mL distilled water and mix thoroughly. Bring to pH 6 to 7 with 6N hydrochloric acid and dilute to 950 mL with distilled water.

Caution: Severe poison, suspected cancer agent.

Phenylarsine oxide titrant standardization

1. Place 200 mL distilled water in the titrating vessel and turn on the stirrer.

2. Add 1 mL of 20 percent sulfuric acid solution.

3. Add 1 mL potassium iodide solution.

4. Carefully add 5.0 mL 0.0025N potassium bi-iodate solution. A pale yellow color should develop.

5. Titrate to the usual amperometric end point (total chlorine).

6. Repeat the standardization process. Duplicate titrations should agree within 0.05 mL.

The normality (N) of the PAO titrant can be determined by the following formula.

$$N \text{ PAO} \times \text{mL PAO} = N \text{ Potassium bi-iodate} \times \text{mL Potassium bi-iodate}$$

$$N \text{ PAO} = \frac{0.0025N \times 5\text{mL}}{\text{mL PAO used}}$$

The acceptable range of the PAO titrant normality is ± 5 percent of theoretical normality (0.00564N). Acceptable range = 0.005358N to 0.005922N. If the normality of the titrant is outside the acceptable range, replace the titrant.

Phosphate buffer solution, pH 7.

1. Dissolve 25.4 g anhydrous potassium dihydrogen phosphate (also called potassium monobasic phosphate) (KH_2PO_4) and 34.1 g anhydrous disodium hydrogen phosphate (also called sodium dibasic phosphate) (Na_2HPO_4) in 800 mL distilled water.

2. Add 2 mL sodium hypochlorite solution that contains 5 percent available chlorine (common household bleach). Stopper and mix thoroughly.

3. Store in a cool, dark place away from sunlight or heat for two days, so the chlorine can react completely with the ammonium contaminants usually present in phosphates.

4. Place the bottle in sunlight, indoors or outdoors, until all chlorine disappears. The time required will vary from one day during summer to one week during winter.

5. When no total chlorine remains, transfer the contents of the bottle to a 1-L graduated cylinder and dilute to the 1 L mark with distilled water. Mix thoroughly by pouring back into the bottle.

6. Filter the solution if any precipitate forms on standing.

Potassium iodide solution.

1. Place 105 mL distilled water in a 250-mL flask and boil for 7 to 10 min. Cover the top of the flask with a clean, small, inverted beaker and allow the water to cool to room temperature. To hasten cooling, place the flask in a bath of cold running water.

2. On a rough balance, weigh 5 g potassium iodide (KI). Transfer to the freshly boiled and cooled distilled water and mix thoroughly.

3. Transfer the solution to an amber, glass-stoppered bottle. Store in a dark, cool place, preferably a refrigerator. Discard the solution when a yellow color develops.

Acetate buffer solution, pH 4.

1. Measure 400 mL distilled water into a 1.5-L beaker. Prepare buffer solution in a fume hood.

2. With a 1-L graduated cylinder, measure 480 mL concentrated acetic acid (CH_3COOH) (also called glacial acetic acid) and add to the 400 mL distilled water. Mix thoroughly.

3. Weigh 243 g sodium acetate trihydrate ($NaC_2H_3O_2 \cdot 3H_2O$) and dissolve in the acetic acid solution.

4. Transfer the solution to the 1-L graduated cylinder; dilute to the 1 L mark with distilled water and mix thoroughly.

Distilled water. Use reagent-grade, deionized distilled water, which should be chlorine free.

Potassium bi-iodate, $KH(IO_3)_2$. This reagent is available commercially from major distributors at $0.025N$ and must be diluted using volumetric glassware to $0.0025N$. Dilute 10 mL $0.025N$ potassium bi-iodate with distilled water to a final volume of 100 mL. The solution must be made fresh for each standardization.

1. To prepare the reagent in the laboratory, dry 2–4 g of reagent-grade potassium bi-iodate for 2 hr at $105°C$.

2. Desiccate to room temperature.

3. Dissolve 1.6245 g potassium bi-iodate in distilled water. Using a 500-mL volumetric flask, dilute to a final volume of 500 mL. This is a $0.1N$ solution.

4. Using volumetric glassware, dilute 25 mL $0.1N$ solution with distilled water to a final volume of 1 L. This is a $0.0025N$ solution and must be made fresh for each standardization.

Sulfuric acid, H_2SO_4, 20 percent solution.

Free chlorine determination procedure.

1. Fill the buret with phenylarsine oxide titrant. Record the liquid level in the buret by reading at the bottom of the meniscus. (Fig. 1-17) Guard against a leaky stopcock, which can result in the loss of titrant.

2. Select the sample volume. Measure the sample and distilled water volumes for the indicated residual chlorine ranges.

   If the residual chlorine falls within the range of 0.0 to 4.0 mg/L, use 100-mL sample and distilled water. First, mix 5 mL phosphate buffer solution and 5 mL DPD reagent in a 250-mL flask. Add 100 mL sample and mix. (If the sample is added before buffer, the test does not work.)

   If the residual chlorine exceeds 4.0 mg/L, reduce the sample size and dilute the sample with distilled water to a total volume of 100 mL (Table 7-2). First, mix 5 mL phosphate buffer solution and 5 mL DPD reagent in a 250-mL flask. Add the specified volume of chlorine free water. Add the sample and mix.

3. Unless sample pH is known to be between 6.5 and 7.5, add 1 mL pH 7 phosphate buffer solution.

Table 7-2   Dilution table for various strengths of residual chlorine

Residual Chlorine Range mg/L	Original Sample Volume mL	Distilled Water Volume mL
0.0–4.0	100	0
4.1–8.0	50	50
8.1–16.0	25	75

4. Titrate with standard phenylarsine oxide titrant, watching the needle movement on the microammeter scale. When the needle moves to the 0 end of the scale, return to mid-scale with the proper adjustment for easier observation and greater sensitivity. As the needle activity diminishes, add progressively smaller increments of titrant. Make successive buret readings when the needle action becomes sluggish, signaling the approach of the end point. Subtract the last very small increment that causes no needle response because of overtitration.

5. Read the new buret level at the bottom of the meniscus and calculate the volume of titrant used by subtracting the initial buret reading (step 1) from the end point reading.

6. Calculate free available chlorine by multiplying the result found in step 5 by the appropriate factor (Table 7-3). (Use distilled water to bring the sample volume to 200 mL.)

Combined chlorine determination procedure.

7. To the sample remaining from free chlorine titration, add exactly 1 mL potassium iodide solution.

8. Add 1 mL acetate buffer solution to the sample.

9. Do not refill buret, but continue titration after recording figure for free chlorine. Repeat the titration procedure described in step 4.

10. Read the new buret level at the bottom of the meniscus and record the total volume of titrant used in both the free available chlorine titration and the combined available chlorine titration. This figure represents total chlorine. Multiply this total by the factor given in step 6.

11. Subtract the value in step 6 (free available chlorine) from the value in step 10 (total chlorine) to obtain the combined available chlorine.

Caution: Wash the electrodes, stirrer, and sample container thoroughly to remove every trace of iodide from the apparatus before making the next free available chlorine determination. Confirm complete iodide removal by duplicating the sample.

Chlorine (Residual) Titrimetric Method

DPD (*N,N*-diethyl-*p*-phenylenediamine) is used as an indicator in the titrimetric procedure with ferrous ammonium sulfate [$Fe(NH_4)_2(SO_4)_2 \cdot 6H_2O$]. This simplified procedure is used to determine free, combined, or total chlorine.

Table 7-3 Calculating free available chlorine from amperometric titration results

Sample Volume mL	Multiply mL of Titrant Used by
200	1
100	2
50	4

Warnings/Cautions.

Sample pH must be 6.2 to 6.5 for accurate results. A lower pH enables chloramine to appear as free chlorine. A higher pH causes dissolved oxygen to give a pink color identical to that produced by chlorine. High temperature enables chloramine to appear as free chlorine and increases color fading. Complete the measurements rapidly at high temperatures.

Oxidized manganese that is naturally occurring or added during water treatment as potassium permanganate reacts with DPD to give a pink color identical to that produced by chlorine. In such a case, a correction must be made for this interference. (See *Standard Methods* for further discussion.)

Chlorine dioxide, if present, appears with free chlorine to the extent of one-fifth the total chlorine content. (See *Standard Methods* for further discussion.)

Monochloramine, if present in high concentration, interferes in the free chlorine determination after 1 min of developing time. Therefore, all readings must be made within the specified time interval.

Apparatus.

- a 10-mL buret and support

- a 100-mL graduated cylinder or volumetric pipets for measuring the sample

- one or more 250-mL flasks

- two 5-mL pipets for dispensing DPD reagent and phosphate buffer solution

- a spatula for dispensing potassium iodide crystals

- a dropping pipet or medicine dropper for dispensing sodium arsenite (also called sodium meta-arsenite) ($NaAsO_2$) solution

Reagents.

Distilled water. Use reagent-grade, deionized distilled water, which should be chlorine free.

Dilute sulfuric acid solution.

1. Using a 50-mL graduated cylinder, measure 30 mL distilled water and pour into a 100-mL beaker.

2. With a 10-mL pipet, measure 10 mL concentrated sulfuric acid, H_2SO_4.

3. While stirring, slowly and cautiously add 10 mL sulfuric acid to 30 mL distilled water. Considerable heat is generated by mixing acid and water, so pour slowly and mix well to avoid dangerous spattering. Cool to room temperature before use.

Ferrous ammonium sulfate titrant.

1. Measure 1,200 mL distilled water into a 2-L flask and boil for 5 min. Cover the top of the flask with an inverted, 400-mL beaker and allow to cool to room temperature. Place in a cold water bath to speed cooling.

2. Pour half the freshly boiled and cooled distilled water into a 1.5-L beaker. Add 1 mL dilute sulfuric acid solution and mix.

3. On an analytical balance, weigh 1.106 g reagent-grade ferrous ammonium sulfate [$Fe(NH_4)_2(SO_4)_2 \cdot 6H_2O$]. Carefully transfer to the 1.5-L beaker and dissolve in the distilled water prepared in step 2.

4. Transfer the solution to a 1-L volumetric flask. Rinse the beaker three times with 100-mL portions of distilled water. Add rinsings to flask. Dilute to the mark with distilled water, stopper, and mix thoroughly. 1 mL of this titrant is equivalent to 1 mg/L of chlorine in the titration procedure.

5. Store the titrant in an amber, glass-stoppered bottle away from bright light. Discard after 1 month.

N,N-diethyl-p-phenylenediamine reagent

1. Place 600 mL distilled water in a 1.5-L beaker. Add 8 mL dilute sulfuric acid solution and mix.

2. Weigh 0.2 g disodium ethylenediaminetetraacetate dihydrate (also called ethylenedinitrilotetraacetic acid sodium salt, or EDTA). Carefully transfer to the 1.5-L beaker and dissolve by mixing in solution described in step 1.

3. Weigh either 1 g *N,N*-diethyl-*p*-phenylenediamine oxalate or 1.5 g *p*-amino-*N*, *N*-diethylaniline sulfate. Carefully transfer to the 1.5-L beaker and dissolve by mixing in solution described in step 2.

4. Transfer the combined solution (step 3) to a 1-L graduated cylinder and dilute to the 1 L mark with distilled water. Mix thoroughly by pouring back into the beaker and stirring.

5. Store the reagent solution in an amber, glass-stoppered bottle away from bright light. Discard when the solution becomes discolored.

Caution: The oxalate reagent is toxic. Do not ingest. Dispense the solution with an automatic, safety, or bulb-operated pipet.

Phosphate buffer solution.

1. Weigh the following dry chemicals separately: (1) 24 g disodium hydrogen phosphate, also called sodium dibasic phosphate (Na_2HPO_4); and (2) 46 g potassium dihydrogen phosphate, also called potassium monobasic phosphate (KH_2PO_4).

2. Transfer the weighed chemicals to a 1.5-L beaker and dissolve in 600 mL distilled water. If necessary, heat the solution gently and stir to bring all the chemicals into solution. If heat is used to dissolve the chemicals, cool the solution to room temperature.

3. Weigh 0.8 g EDTA and dissolve in 100 mL distilled water. Add to solution (step 2) and mix.

4. Transfer the mixed solution (step 3) to a 1-L graduated cylinder and dilute to the 1 L mark with distilled water. Mix thoroughly by pouring back into the beaker and stirring.

5. Weigh 20 mg mercuric chloride ($HgCl_2$) and add to solution (step 4) to prevent mold growth and interference in the free available chlorine test caused by any trace of iodide in the reagents.

Caution: Mercuric chloride is toxic. Do not ingest.

6. Potassium iodide crystals.

7. Sodium arsenite solution (to estimate manganese interference). Weigh 5 g sodium arsenite (also called sodium meta-arsenite) ($NaAsO_2$). Dissolve in 1 L distilled water.

Caution: Poison. Handle with extreme caution. Do not ingest. Dispense the solution with an automatic, safety, or bulb-operated pipet.

Free available chlorine determination procedure.

1. Fill the buret with ferrous ammonium sulfate [$Fe(NH_4)_2(SO_4)_2 \cdot 6H_2O$] (FAS). Record the liquid level in the buret by reading the bottom of the meniscus. Guard against a leaky stopcock, which can result in the loss of titrant.

2. Select the sample volume. Measure the sample and distilled water volumes for the indicated residual chlorine ranges.

    If the residual chlorine falls within the range of 0.0 to 4.0 mg/L, use 100-mL sample and no distilled water. First, mix 5 mL phosphate buffer solution and 5 mL DPD reagent in a 250-mL flask. Add 100-mL sample and mix. (If the sample is added before the buffer, the test does not work.)

    If the residual chlorine exceeds 4.0 mg/L, reduce the sample size and dilute the sample with distilled water to a total volume of 100 mL (Table 7-2). First, mix 5 mL phosphate buffer solution and 5 mL DPD reagent in a 250-mL flask. Add the specified volume of distilled water. Add the sample and mix.

3. If the sample turns pink or red, add FAS titrant from the buret and swirl the flask constantly until the pink just disappears.

4. Read the new buret level at the bottom of the meniscus and calculate the volume of titrant used by subtracting the initial buret reading (step 1) from the present reading.

5. Calculate the free available chlorine by multiplying the result found in step 4 by the appropriate factor. See Table 7-4.

Combined chlorine determination procedure.

6. Add several potassium iodide (KI) crystals (total weight 0.5 to 1.9 g) to the flask from step 5 (free available chlorine) and mix to dissolve. Let the solution stand for 2 min so the chloramine in the sample can convert the iodide to iodine as evidenced by the return of the pink or red color.

Table 7-4   Calculating free available chlorine from titration results

Original Sample Volume mL	Multiply mL Titrant Used by
100	1
50	2
25	4

7. Resume titrating with small volumes of FAS titrant until the pink or red color again disappears.

8. Read the new buret level at the bottom of the meniscus and record the total volume of titrant used in both the free available chlorine titration (step 4) and the combined chlorine titration (step 7). Multiply this total by the factor given in step 5.

9. Subtract the value in step 5 from the value in step 8 (combined chlorine) to obtain the value for combined available chlorine.

Estimating manganese interference procedure.

1. Place 5 mL phosphate buffer solution, 1 small crystal of potassium iodide, and 0.5 mL sodium arsenite solution in a 250-mL flask and mix.

2. Add 100-mL sample and mix.

3. Add 5 mL DPD reagent and mix.

4. If the solution turns pink or red, manganese interference is present. Titrate with FAS titrant until the pink disappears.

5. Read the new buret level at the bottom of the meniscus and calculate the volume of titrant used by subtracting the initial buret reading from the present reading. Multiply the result by the factor given in step 5 to obtain the manganese interference.

6. Subtract manganese interference from the results in step 5 (free chlorine) and 8 (combined chlorine) to obtain the true, free available chlorine and total available chlorine, respectively.

7. If the values on the two spiked sample portions are higher by the amount that was artificially added, the result on the original unspiked sample can be assumed to be correct. If the recoveries exceed or fall short of the calculated amount by more than experimental error, the trouble may be attributed to an interference in the unknown sample.

This page intentionally blank

Chapter **8**

Chlorine Demand

PURPOSE OF TEST

Chlorine is added to a water supply to ensure its bacteriological acceptability or to improve its chemical, physical, or taste and odor characteristics.

This procedure is suitable for determining the approximate amount of chlorine needed to produce a chlorine residual in source waters of drinking water quality that contain comparatively little pollution. The bacteriological safety of a water is, in most cases, ensured when a slight excess of chlorine is present.

LIST OF SIMPLIFIED METHODS

Refer to *Standard Methods for the Examination of Water and Wastewater*, Section 2350 B. Chlorine Demand/Requirement.

SIMPLIFIED PROCEDURE

Chlorine Demand

Chlorine demand varies with the amount applied, time of contact, pH, and temperature. Some chemicals, such as ammonia, hydrogen sulfide, and iron increase demand on chlorine. For comparative purposes, all test conditions should be listed, including the method of determining the residual chlorine.

Warnings/Cautions.

All chlorine solutions and samples must be kept away from direct sunlight and chlorine-consuming fumes, such as ammonia and sulfur dioxide.

The chlorine solution should be standardized, or the results can only be approximate. For more information on preparing a standard curve, see the Standard Calibration Curves section in chapter 1.

If the test is bacteriologic, all glassware should be thoroughly cleaned and sterilized.

If the test is to measure taste and odor with and without chlorine, make sure the water is safe to drink. This test does not ensure that the water is disinfected.

Apparatus.

In addition to the apparatus described in the section on chlorine determination (free or residual) in chapter 7, the following are also needed.

- five or more clean quart-size (1-L) bottles or flasks

- a dropping pipet or medicine dropper for dispensing the chlorine dosing solution that delivers 1 mL in 20 drops or 50 μL per drop. To use the dropper, hold vertically and let the drops form slowly. Set this dropper aside for exclusive use in this procedure.

Reagents.

In addition to the reagents described in the section on chlorine determination (free or residual) in chapter 7, the following are needed:

Stock chlorine solution. Purchase a bottle of household bleach. This product should contain approximately 5 percent available chlorine, representing 52,500 mg/L. Store in a cool dark place, such as a refrigerator, and ensure that the container is tightly sealed to maintain the chemical strength. Replenish once or twice a year.

Prepare two dosing solutions. The strong solution is prepared by adding 20 mL of the stock chlorine solution to 80 mL distilled water. Use separate graduated cylinders to measure each volume. Keep the stock chlorine solution away from the distilled water source to prevent contamination of the distilled water by chlorine fumes. Mix well and stopper in a clean brown bottle. Store in a refrigerator to ensure chemical strength for at least a month. The strong solution is a 10 mg/mL chlorine solution. Each drop (using the dropper described previously) of this solution in 500 mL water sample represents a chlorine dose of 1 mg/L.

The dilute solution is prepared by adding 10 mL strong solution to 90 mL distilled water. Use separate graduated cylinders. Mix well and stopper. Store in a brown bottle and place in a refrigerator for as long as a week. This solution is a 1 mg/mL chlorine solution. Each drop (using the dropper described previously) of this solution in 500 mL water sample represents a chlorine dose of 0.1 mg/L.

The chlorine strength of these solutions may be determined by one of the methods described in the section on residual chlorine in chapter 7.

Procedures.

1. Add 500 mL water sample to 5 to 10 bottles or flasks. A graduated cylinder is accurate enough.

2. Bring samples to room temperature or to the temperature of the water in the treatment plant.

3. Add the number of drops listed in Table 8-1 of the dilute chlorine solution to the separate bottles or flasks.

   Select and prepare a smaller number of samples from this series. Set up the dosing schedule so the first sample shows no residual chlorine at the end of the desired contact time, which may be 1, 2, 4, or even 24 hr.

4. Mix the contents of each bottle or flask and set aside at the desired temperature (see step 2). Keep the samples in the dark to avoid chlorine degradation.

5. Choose contact time. At the end of the selected time, mix the sample in the first bottle and determine the residual chlorine (see Residual Chlorine Method in chapter 7). Free available chlorine can be determined with the same method.

6. Record the chlorine dose added, the total or free available residual chlorine found, the contact time, and the temperature of the sample during the holding time.

7. Repeat steps 5 and 6 for each sample.

8. Calculate the chlorine demand in each case by subtracting the residual chlorine from the dose added.

9. If no residual chlorine is found in the sample dosed with 2 mg/L chlorine (the highest level), use the strong chlorine dosing solution to prepare the series of samples provided in Table 8-2.

10. Complete the determination as described in steps 4–8.

Table 8-1   Dilutions for determining chlorine demand using dilute solution

Drops of Dilute Solution	Chlorine mg/L
2	0.2
4	0.4
6	0.6
8	0.8
10	1.0
12	1.2
14	1.4
16	1.6
18	1.8
20	2.0

Table 8-2   Dilutions for determining chlorine demand using strong solution

Drops of Strong Solution	Chlorine mg/L
2	2
4	4
6	6
8	8
10	10

This page intentionally blank

Chapter **9**

Chlorine Dioxide

PURPOSE OF TEST

Chlorine dioxide can be used to control taste and odors in drinking water, especially those caused by phenolic compounds, to oxidize and remove iron and manganese, and to disinfect. Treatment plants also add chlorine dioxide near the beginning of the treatment system to delay adding chlorine until later in the process. By delaying the addition and lowering the amount of chlorine added, plants may be able to lower the concentrations of trihalomethanes formed in the distribution system.

Chlorine dioxide is a volatile gas that is usually generated at the point of use by reacting strong chlorine and sodium chlorite solutions. The chlorine dioxide solution is deep yellow and should be handled with care. Concentrated chlorine dioxide solutions and waters treated with chlorine dioxide may contain a variety of chlorine species, including chlorine dioxide (ClO_2), free chlorine (Cl_2), combined chlorine (NH_2Cl), chlorite (ClO_2^-), and chlorate (ClO_3^-).

Analytical measurement methods should be able to distinguish between these species. The concentrated chlorine dioxide feed solution may be analyzed to provide information on generator efficiency. Analysis of treated waters is important to determine residual chlorine dioxide levels and to monitor the concentrations of inorganic by-products chlorite and chlorate, which are suspected of causing hemolytic anemia in humans; thus, their concentrations should be kept as low as possible.

LIST OF SIMPLIFIED METHODS

The large number of chlorine dioxide/chlorine species and interactions between these species makes their measurement difficult, especially at low concentrations. Two basic types of methods available and recognized by the Safe Drinking Water Act for analyzing these species are *N,N*-diethyl-*p*-phenylenediamine (DPD) colorimetric methods and amperometric titration methods. Appendix A provides a list of manufacturers and suppliers.

DPD methods include commercial comparator kits (color wheels) and a titrimetric method that uses ferrous ammonium sulfate [$Fe(NH_4)_2(SO_4)_2 \cdot 6H_2O$] (FAS) as the titrant. The color wheel measures only the total chlorine dioxide species concentration; the DPD titrimetric method can measure individual chlorine dioxide, free chlorine, combined chlorine, and chlorite concentrations. Some errors have been associated with the DPD method. For this reason, the method will be omitted from the next (21st) edition of *Standard Methods*. This method is included for general reference.

The amperometric titration method measures chlorine dioxide, total chlorine, chlorite, and chlorate. Because of its extreme difficulty, the chlorate portion of the procedure is not included here. DPD methods are easier to perform but are less accurate than amperometric methods. However, the amperometric methods require specialized equipment and considerable analytical skill.

Ion chromatographic methods are also available for measuring chlorite and chlorate species. For more information on chlorine dioxide analyses, see the references section for this chapter.

SIMPLIFIED PROCEDURES

General Information

Warnings/Cautions.

Obtain and handle all samples to prevent aeration. Chlorine dioxide is a volatile gas and may escape from the sample. Seal sample containers with a minimum of headspace. Transport field samples to the laboratory in a cooler that contains ice. Analyze samples as soon as possible.

When making dilutions of concentrated chlorine dioxide solutions, use a volumetric pipet to carefully remove a portion of the solution. Drain the pipet tip by placing the tip under the surface of the dilution water.

Keep all samples away from sunlight, which rapidly decomposes chlorine dioxide.

DPD Color Wheels

See General Information in this section.

Warnings/Cautions.

Exposure to direct sunlight may cause color wheels to fade. Rinse viewing tubes thoroughly between sample analyses.

Apparatus, Reagents, and Procedure.

Color wheels with the required reagents and instructions can be purchased commercially (see appendix A).

DPD Titrimetric Method

Reading the DPD titrimetric method described for Chlorine (Residual)—Titrimetric Method in chapter 7 may be helpful.

Warnings/Cautions.

Obtain and handle all samples so minimal aeration occurs. Chlorine dioxide is a volatile gas and may escape from the sample. Seal sample containers with a minimum of headspace. Transport field samples to the laboratory in a cooler that contains ice. Analyze samples as soon as possible.

When making dilutions of concentrated chlorine dioxide solutions, use a volumetric pipet to carefully remove a portion of the solution. Drain the pipet tip by placing the tip under the surface of the dilution water.

Keep all samples out of sunlight, which rapidly decomposes chlorine dioxide.

The end point of DPD titrations for free chlorine occurs immediately after all color has disappeared for the first time. As time passes, the red or pink may reappear as chloramines form and react with the DPD color reagent, causing a false increase in apparent free chlorine.

The interference by manganese described for chlorine (general) also applies to chlorine dioxide measurements. Oxidized manganese, whether present naturally or as potassium permanganate ($KMnO_4$) added during treatment, reacts with DPD to produce a red or pink color identical to that produced by chlorine. In such cases, a correction must be applied.

Iron may also interfere. Avoid this problem by adding additional ethylene-diamine tetraacetic acid (EDTA), disodium salt.

Apparatus.

- equipment listed for the analysis of Chlorine (Residual)—Titrimetric Method in chapter 7

- a bottle and medicine dropper for adding glycine solution

- a 5-mL pipet for adding sulfuric acid

Reagents.

All reagents listed for the analysis of Chlorine (Residual)—Titrimetric Method in chapter 7.

Glycine Solution. Dissolve 10 g glycine (NH_2CH_2COOH) in 100 mL distilled water.

Sulfuric Acid Solution, 1.8N. Dilute 5 mL concentrated sulfuric acid (H_2SO_4) to 100 mL with distilled water.

Sodium Bicarbonate Solution. Dissolve 27.5 g sodium bicarbonate ($NaHCO_3$) in 500 mL distilled water.

Chlorine dioxide determination procedure.

1. Measure 100 mL of sample using a graduated cylinder and place the sample in a 250-mL Erlenmeyer flask. Add 2 mL glycine solution.

2. Place 5 mL buffer reagent and 5 mL DPD solution into a separate 250-mL Erlenmeyer flask and mix. Add approximately 200 mg EDTA, disodium salt.

3. Add the glycine-treated sample to the flask prepared in the previous step and mix.

4. Quickly titrate the sample with the FAS titrant until the red or pink color just disappears.

5. Record the volume of titrant added. This is the value *A*.

Free chlorine determination procedure.

1. Place 5 mL buffer reagent and 5 mL DPD solution into a 250-mL Erlenmeyer flask and mix. Add approximately 200 mg of EDTA, disodium salt.

2. Measure 100 mL sample with a graduated cylinder, and add to the flask prepared in the previous step, and mix.

3. Quickly titrate the sample with the FAS titrant until the red or pink just disappears.

4. Record the volume of titrant added. This is the value B.

Combined chlorine determination procedure.

1. Immediately after measuring the free chlorine by obtaining value B previously listed for free chlorine, add approximately 1 g potassium iodide (KI) crystals to the same flask and mix. Let sample stand for 2 min.

2. Titrate with FAS titrant until the red or pink color disappears. If the combined chlorine concentration is greater than 1 mg/L, the red or pink may return because of an incomplete reaction. If this occurs, let the sample stand 2 more min and then continue titrating the sample until the color just disappears.

3. Record the volume of titrant added. This is value C.

Total available chlorine (including chlorite) determination procedure.

1. Immediately after measuring the combined chlorine by obtaining the value C previously listed for combined chlorine, add 1 mL 1.8N sulfuric acid solution to the same flask and mix. Let sample stand for 2 min.

2. Add 5 mL sodium bicarbonate solution and mix.

3. Titrate with FAS titrant until the red or pink color just disappears.

4. Record the volume of titrant added. This is value D.

Estimation of manganese interference determination procedure.

1. Place 5 mL phosphate buffer, one small potassium iodide crystal, and 0.5 mL sodium arsenite solution into a 250-mL Erlenmeyer flask and mix.

2. Add 100 mL sample and mix.

3. Add 5 mL DPD reagent and mix.

4. If the solution turns red or pink color, manganese interference is present. Titrate with FAS titrant until the red or pink color just disappears.

5. Record the volume of titrant added. This is value E.

Calculate results. All units are mg/L.

- chlorine dioxide (ClO_2) = $1.9 \times A$

- free chlorine (Cl_2) = $B - A$

- combined chlorine (NH_2Cl) = $C - B$
- chlorite (ClO_2^-) = $D - [C + (4 \times A)]$

If required, subtract value E from each of the previously calculated results to correct for manganese interference.

Amperometric Titration Method

Read the amperometric titration method described for chlorine (residual) in chapter 7 to become familiar with amperometric titration end points.

Warnings/Cautions.

Obtain and handle all samples so minimal aeration occurs. Chlorine dioxide is a volatile gas and may escape from the sample. Seal sample containers with a minimum of headspace. If possible, transport field samples to the laboratory in a cooler that contains ice. Analyze samples as soon as possible.

When making dilutions of concentrated chlorine dioxide solutions, use a volumetric pipet to carefully remove a portion of the solution. Drain the pipet tip by placing the tip under the surface of the dilution water.

Keep all samples out of sunlight, which rapidly decomposes chlorine dioxide.

Interferences by manganese, copper, and nitrite are possible during the analysis for chlorite.

Apparatus.

- amperometric titrator with a platinum–platinum electrode
  Refer to the section on Chlorine (Residual)—Amperometric Method in chapter 7 for the use and care of amperometric titration units.

- a 250-mL beaker for the titration vessel
  Some amperometric titrators provide a special titration cup to hold the sample.

- a 250-mL graduated cylinder for sample measurement

- a 500-mL graduated cylinder for the sample container when purging with nitrogen gas

- a 1-mL pipet for adding phosphate buffer

- a 2-mL pipet for adding hydrochloric acid

- a purging unit that consists of a scrubber vessel with a 5 percent KI solution, connective tubing, and a gas purging unit (see Figure 9-1)

Reagents.

Standard sodium thiosulfate solution, $Na_2S_2O_3 \cdot 5H_2O$, 0.1N or standard phenylarsine oxide (C_6H_5AsO), 0.00564N. These can be purchased commercially or prepared in the laboratory as described in the section on Chlorine (Residual)—Amperometric Method in chapter 7.

Phosphate buffer solution. Dissolve 25.4 g anhydrous potassium dihydrogen phosphate (KH_2PO_4) and 34.1 g anhydrous disodium hydrogen phosphate, also called sodium dibasic phosphate ($NaHPO_4$) in 1 L distilled water.

Potassium iodide (KI) crystals.

Hydrochloric acid, 2.5N. Cautiously add 200 mL concentrated hydrochloric acid (HCl) to 500 mL distilled water. Mix and dilute to 1 L.

Nitrogen purge gas.

Potassium iodide scrubber solution, KI, 5 percent. Dissolve 5 g potassium iodide in 100 mL distilled water. Discard solution and replace with fresh solution at the first sign of yellow discoloration.

Total chlorine and chlorine dioxide determination procedure.

1. Place 1 mL phosphate buffer in the titration vessel.

2. Measure 200 mL sample (or a portion of the sample diluted to 200 mL) using a graduated cylinder and add to the titration vessel. Pour sample slowly with a minimum of aeration to avoid losing chlorine dioxide.

3. Begin to stir the sample and add 1 g potassium iodide crystals.

4. Titrate to the end point using the sodium thiosulfate or phenylarsine oxide reagent.

5. Record the volume of titrant added. The value A = mL titrant/mL of sample.

Chlorine dioxide and chlorite determination procedure.

1. Immediately after completing step 1 for total chlorine and chlorine dioxide, add 2 mL 2.5N hydrochloric acid to the same sample. Cover titration vessel with a box and let the sample stand in the dark for 5 min.

2. Remove sample from the dark and titrate to the end point using the sodium thiosulfate or phenylarsine oxide reagent.

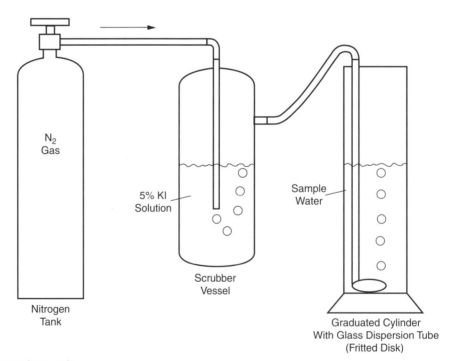

Figure 9-1   Gas purging unit

3. Record the volume of titrant added. The value B = mL titrant/mL sample.

Nonvolatilized chlorine determination procedure.

1. Place 1 mL phosphate buffer into the purge vessel.

2. Add 200 mL fresh sample (or a portion of the sample diluted to 200 mL) to the purge vessel.

3. Purge with nitrogen gas for 20 min. The gas flow should be high enough to create a fairly turbulent flow of gas bubbles from the gas dispersion tube. However, it should not be so high that potassium iodide solution is carried over from the scrubber vessel to the sample.

4. Rinse the sides of the purge vessel with a small amount of distilled water and transfer the sample to the titration vessel.

5. Begin to stir the sample and add 1 g potassium iodide crystals.

6. Titrate to the end point using the sodium thiosulfate or phenylarsine oxide reagent.

7. Record the volume of titrant added. The value C = mL titrant/mL sample.

Chlorite procedure.

1. Immediately after completing step 3 for nonvolatilized chlorine, add 2 mL of 2.5 N hydrochloric acid to the same sample. Cover the titration vessel with a box and let the sample stand in the dark for 5 min.

2. Remove the sample from the dark and titrate to the end point using the sodium thiosulfate or phenylarsine oxide reagent.

3. Record the volume of titrant added. The value D = mL titrant/mL sample.

Calculate results. (All units are mg/L.)

- $F = B - D$
- chlorine dioxide (ClO_2) = $(5/4) \times F \times N \times 13490$
- chlorine (Cl_2) = $[A - (F/4)] \times N \times 35453$
- chlorite (ClO_2^-) = $D \times N \times 16863$

This page intentionally blank

Chapter **10**

Color

PURPOSE OF TEST

Color in water may result from any or all of the following: natural metallic ions, such as iron or manganese; humus or peat materials; plankton; weeds; or industrial waste. The term *color* means true color. The most common colors in source water are yellow and brown. Suspended materials can add an apparent color.

Color is removed to make a water suitable for general and industrial applications. On a plant level, true color is often removed from a colored soft water by coagulating with alum in the pH acid range (see chapter 20 for pH method). Free residual chlorination and superchlorination may help reduce color.

This procedure is designed to measure the true color in water and is useful for plants that must treat a colored supply.

LIST OF SIMPLIFIED METHODS

Refer to *Standard Methods for the Examination of Water and Wastewater* for a discussion of two additional methods that are used when interferences occur:

- Section 2120 C. Spectrophotometric Method
- Section 2120 E. ADMI Tristimulus Filter Method

SIMPLIFIED PROCEDURES

Visual Comparison Method

Warnings/Cautions.

Even a slight amount of turbidity can interfere with this determination and cause the color to appear greater than the true color. Filtration is not recommended because it may remove some color. *Standard Methods* lists two possible ways to avoid this interference.

Color is highly dependent on the pH of the sample. When reporting a color value, specify the pH at which the color is determined.

Potassium chloroplatinate and cobaltous chloride are the two reagents used to make the color standards. If potassium chloroplatinate is not available, use chloroplatinic acid prepared from metallic platinum. Commercial chloroplatinic acid is hygroscopic and may vary in platinum content.

Apparatus.

- Nessler tubes, matched, tall form, 50-mL capacity

- pH meter (see pH method in chapter 21)

- burets, as needed

- clean stoppers for each Nessler tube

Reagents.

Stock color standard

1. Weigh 1.246 g potassium chloroplatinate (K_2PtCl_6) in glass. Transfer to a 250-mL glass beaker.

2. Weigh 1.0 g cobaltous chloride ($CoCl_2 \cdot 6H_2O$) in glass. Transfer to the same 250-mL glass beaker.

3. Add 50 mL distilled water and dissolve the two powders.

4. Add 100 mL concentrated hydrochloric acid (HCl) and mix thoroughly. Always add acid to water, not water to acid.

5. Transfer this solution to a 1-L volumetric flask. Rinse the beaker with three volumes distilled water, adding each to the flask. Dilute to the 1-L mark. (This solution is equivalent to 500 color units.)

Procedure.

1. Prepare the following color standards by adding the volumes listed in Table 10-1 to separate Nessler tubes. For more information on preparing a standard curve, see the Standard Calibration Curves section in chapter 1.

2. Add distilled water to the 50-mL mark and mix. Protect these standards by capping with clean rubber stoppers. They may be kept for several months in this manner. They will keep indefinitely if the tubes are permanently sealed with clear glass caps.

Color determination for samples less than 70 color units

1. Place 50 mL clear sample in a separate Nessler tube. Compare color with those of the standards by looking vertically down through the tubes toward a white specular surface placed so light is reflected upward through the columns of liquid.

2. If color exceeds 70 units, dilute the sample with distilled water in known proportions until the sample is within the range of the standards. Multiply the observed color units by the dilution factor. Report this value as total color.

If turbidity is present, it may be removed by centrifuging or as determined by one of the other simplified methods listed earlier in this chapter.

NOTE: If the color is between the prepared standards, make an educated guess. Look at the intensity of the sample color. When the color of the sample has a brighter intensity than the color of a standard, that color is somewhere between that standard and the next higher one. Estimate the color of the sample from the closest standard. For example: Sample color is between 10 and 20, but much closer to 10 than 20, so the approximate sample color is 12 or 13 units.

Table 10-1   Dilutions to prepare color standards

Color Solution mL	Color Units
0	0
0.5	5
1	10
2	20
3	30
4	40
5	50
6	60
7	70

This page intentionally blank

Chapter **11**

Conductivity

PURPOSE OF TEST

Conductivity of drinking water is often related to the concentration of dissolved mineral salts or filterable residue. Departures from normal conductivity may signal changes in the mineral composition of the source water, seasonal variations in reservoirs, daily chemical fluctuations in rivers, or the intrusion of industrial wastes. Only experience with a given water supply will confirm the reasons for changes in conductivity.

Measuring conductivity can help determine the amount of ionic reagent needed to affect precipitation or neutralization reactions in water treatment. This is accomplished by plotting conductivity against buret readings, with the end point noted by a change in slope of the curve. Conductivity also can offer a clue to the size of an unknown water sample that can be taken for a common chemical analysis.

Conductance is the reverse of resistance where resistance is expressed in ohms. Conductance measures the ability to conduct a current and is expressed in reciprocal ohms or mhos. A more convenient unit in water analysis is micromhos, which is related to temperature. A standard temperature of 25°C is used for reference. Most drinking waters in the US exhibit a conductivity of 50 to 1,000 micromhos/cm at 25°C. Freshly distilled water has a conductivity of 0.5 to 2 micromhos/cm, increasing to 2 to 4 micromhos/cm after a few weeks because atmospheric carbon dioxide and, sometimes, ammonia is absorbed.

In the International System of Units (SI) the reciprocal of the ohm is the siemens (S). Conductivity is reported as millisiemens per meter (mS/m). 1 mS/m = 10 μmho/cm and 1 μS/cm = 1 μmho/cm. To report results in SI units of mS/m divide μmho/cm by 10.

LIST OF SIMPLIFIED METHODS

Refer to *Standard Methods for the Examination of Water and Wastewater*, Section 2510 Conductivity.

SIMPLIFIED PROCEDURE

Conductivity

Warnings/Cautions.

Conductivity increases with temperature at a rate of approximately 2 percent per °C; thus, accurate temperature measurement is essential.

To minimize cell and electrode fouling, water that contains substantial suspended matter is best settled before conductivity is measured.

Oils, greases, and fats can also coat electrodes and affect the accuracy of readings. This can be overcome by immersing the electrodes in detergent solutions and rinsing well with distilled water afterward.

Apparatus.

- Self-contained conductivity instruments consist of a source of alternating current (AC), a Wheatstone bridge, a null indicator, and a conductivity cell. The null point is disclosed by an AC galvanometer or a cathode-ray tube. Some instruments give readings in ohms; others read directly in conductivity or conductance units of micromhos/cm.

  Newer units are compact, battery-operated, temperature-correcting, direct-reading conductivity meters suitable for field applications. Quality laboratory and field instruments are equipped for operations over a wide conductivity range.

- The conductivity cell makes up one arm of the Wheatstone bridge. The cell contains a pair of rigidly mounted electrodes, the design, shape, size, and position of which influence the numerical value of the cell constant.

  The cell constant is determined by measuring the resistance of a standard $0.01M$ potassium chloride (KCl) solution at 25°C. A cell constant in the 0.1 to 2.0 range is satisfactory for measuring most drinking waters. A cell with a constant of 0.1 yields the best results in the 1 to 400 micromhos/cm range; one with a constant of 2.0 functions best in the 200 to 10,000 micromhos/cm range.

- Conductivity cells that contain platinized electrodes are available in either the pipet or immersion form, suitable for laboratory measurements. Other types of electrodes are widely used for continuous monitoring and field studies.

- A thermometer should cover the range of 23°C to 27°C and be readable to 0.1°C.

Reagents.

Conductivity water. Pass distilled water through a mixed-bed deionizer and discard the first liter. Conductivity should be less than 1 micromho/cm.

Standard potassium chloride solution. This solution can be purchased commercially or prepared in the laboratory. Weigh 745.6 mg of anhydrous potassium chloride (KCl) in weighing dish. Dissolve into 1 L conductivity water at 25°C. This is the standard reference solution, which at 25°C has a conductivity of 1412 micromhos/cm. It is suitable for most samples when the cell has a constant between 1 and 2. Store in a glass-stoppered borosilicate glass bottle.

Procedure.

1. Determine the cell constant. Rinse the conductivity cell with at least three portions of the standard reference potassium chloride $0.01M$ solution. Adjust the temperature of a fourth portion to 25.0 ± 0.1°C. Measure the resistance of this portion and note the temperature.

    Compute the cell constant C by:

    $$C \text{ (cm}^{-1}) = (0.001\ 412)(R_{KCL}) \ [1 + 0.019 \ (t - 25)] \tag{11-1}$$

    Where:

    > C = cell constant
    > R_{KCL}= measured resistance, ohms
    > t = observed temperature, $°C$

2. Each day the conductivity meter is used, determine the cell constant with the standard potassium chloride solution. Repeat during the course of the day if many conductivity readings are to be done.

3. Thoroughly rinse the conductivity cell several times with part of the sample. Extra rinsing is recommended whenever sample conductivities vary by a factor of 5 or more.

4. Adjust the temperature of the sample to be measured to 25 ± 0.1°C. Slowly submerge the electrodes into the water sample to allow the water to rise above the vent holes of the cell. Ensure that no air bubbles form or cling to the measuring surfaces.

5. Determine the temperature of the water sample to the nearest 0.1°C and record.

6. Follow the manufacturer's instructions for operating the instrument and measuring resistance. Record resistance and temperature.

7. Rinse the electrode and repeat steps 4–6.

8. Compute the sample conductivity by the following:

    $$k = \frac{(1,\ 000,\ 000)(C)}{R_m [1 + 0.019(t - 25)]} \tag{11-2}$$

    Where:

    > k = conductivity, μmhos/cm
    > C = cell constant, cm^{-1}
    > R_m = measured resistance of sample, ohms
    > t = temperature of measurement

9. If the conductivity of the sample exceeds the range of the instrument, dilute with conductivity water and repeat steps 4–7.

Report conductivity values below 1,000 micromhos/cm in whole numbers and above 1,000 to three significant figures. When dilution is necessary, report the dilution factor and the reading on the diluted sample as well.

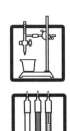

Chapter **12**

Dissolved Oxygen

PURPOSE OF TEST

Dissolved oxygen (DO) is not considered a contaminant in water, but it does play a major role in clean lake studies and reservoir management plans. DO is also an important factor in stream quality standards. Applied by aeration to source water supplies and deep well waters, DO oxidizes and removes dissolved iron and manganese. DO improves the taste and palatability of drinking water, but increases a water's corrosivity.

LIST OF SIMPLIFIED METHODS

Several methods for testing DO are described in *Standard Methods for the Examination of Water and Wastewater*, Section 4500–O. The best methods for testing clean source waters and finished waters for DO are the membrane electrode method (Section 4500–O G.) and azide modification of the Winkler method (Section 4500–O C.).

SIMPLIFIED PROCEDURES

Membrane Electrode Method

Warnings/Cautions.

This method may be used at the sampling site or analyzed later in samples collected in biochemical oxygen demand (BOD) bottles. Prolonged use of membrane electrodes in waters that contain gases, such as hydrogen sulfide (H_2S), tends to lower cell sensitivity. Eliminate this interference by frequently changing and calibrating the membrane electrode.

Because membrane electrodes offer the advantage of analysis on location, they eliminate errors caused by sample handling and storage. If sampling is required, use the same precautions suggested for the azide modification of the Winkler method described in this chapter.

Apparatus.

- a DO meter

- an oxygen-sensitive membrane electrode

Procedure.

1. Prepare the electrode according to the manufacturer's instructions.

2. Calibrate the probe according to the manufacturer's instructions provided with the probe. Use the Winkler method, the saturated water method, or the air method. Altitude and barometric pressure values must be factored into this procedure. Air calibration is highly reliable and much simpler than the other two methods.

3. Place the probe in the sample to be measured.

4. Stir with a submersible stirrer or manually by raising and lowering the probe about one foot per second.

5. Allow enough time for the probe to stabilize before reading the temperature and DO.

6. Read DO concentration from the meter.

Azide Modification of the Winkler Method

Warnings/Cautions.

This method is applicable for most source waters intended for drinking purposes. If a water contains more than 0.1 mg/L nitrate nitrogen, or more than 1 mg/L ferrous iron or other oxidizing or reducing agents, consult *Standard Methods* for more appropriate methods. All reagents for this method are commercially available.

Collect samples from representative locations. Ensure that ambient air is not introduced into the sample during the collection procedure. Consult *Standard Methods* for details regarding special sampling devices. Fix samples to be analyzed by the Winkler method with manganous sulfate solution on the sample site.

Apparatus.

- one or more standard 250- to 300-mL BOD bottles with narrow mouths and glass stoppers

- a 1-L volumetric flask

- one 250-mL graduated cylinder

- one 1.5-L beaker

- one or more wide-mouth 500-mL Erlenmeyer flasks

- a 25-mL buret and support

- four measuring pipets for transferring 1-mL portions of reagents

Reagents.

Manganous sulfate solution.

1. Weigh 480 g manganous sulfate tetrahydrate ($MnSO_4 \cdot 4H_2O$), 400 g manganous sulfate dihydrate ($MnSO_4 \cdot 2H_2O$), or 364 g manganous sulfate monohydrate ($MnSO_4 \cdot H_2O$). This reagent is available commercially.

2. Transfer the manganese salts to a 1.5-L beaker and dissolve in 600 mL distilled water.

3. Dilute to 1 L with distilled water in a 1-L volumetric flask. Mix thoroughly by pouring back into the beaker and stirring.

Alkali-iodide-azide solution.

1. Measure 600 mL distilled water and pour into a 1.5-L beaker. Gradually add 500 g sodium hydroxide (NaOH) pellets, stirring constantly. Surround the beaker with either running water or ice water and work under the fume hood because this solution generates acrid fumes and heat.

2. Weigh 150 g potassium iodide (KI), transfer to a 250-mL beaker, and dissolve in 150 mL distilled water.

3. Add potassium iodide solution to sodium hydroxide solution with constant and thorough mixing. This reagent is available commercially.

4. Transfer the mixed solution to a 1-L volumetric flask and dilute to the 1,000 mL mark with distilled water. Mix thoroughly by pouring back into the beaker and stirring.

5. Weigh 10 g sodium azide (NaN_3), transfer to a 250-mL beaker, and dissolve in 40 mL distilled water.

6. Add sodium azide to the mixed solution with constant and thorough mixing. Store in a bottle that has a rubber or plastic stopper.

Sulfuric acid, concentrated, H_2SO_4.

Starch indicator solution.

1. Use either an aqueous solution or soluble starch powder. To make an aqueous solution, dissolve 2 g laboratory-grade soluble starch and 0.2 g salicyclic acid in 100 mL hot distilled water.

Boiled distilled water (CO_2 free).

1. Transfer at least 2 L distilled or deionized water into a large unstoppered flask or bottle and boil for at least 15 min to expel any CO_2, then cover the flask or bottle and cool to room temperature in a running cold water bath.

2. Prepare the boiled distilled water immediately before it is needed to prepare the following sodium thiosulfate titrant.

Sodium thiosulfate titrant, 0.0250N.

1. Dissolve 6.205 g sodium thiosulfate ($Na_2S_2O_3 \cdot 5H_2O$) in distilled water. Add 1.5 mL 6N NaOH or 0.4 g solid NaOH and dilute to 1,000 mL with distilled water. Standardize with bi-iodate solution.

2. As an alternate to sodium thiosulfate, PAO may be used.

Standardization

1. Dissolve approximately 2 g KI, free from iodate, in an Erlenmeyer flask with 100 to 150 mL distilled water. Add 1 mL $6N$ H_2SO_4 or a few drops of concentrated H_2SO_4 and 20.00 mL standard bi-iodate solution. Dilute to 200 mL and titrate liberated iodine with thiosulfate titrant, adding starch toward end of titration, until a pale straw color is reached. When the solutions are of equal strength, 20.00 mL 0.025M $Na_2S_2O_3$ should be required. If not, adjust the $Na_2S_2O_3$ solution to 0.025M.

Procedure.

1. Collect the sample as described previously. Use a narrow-mouth, glass-stoppered BOD bottle of 250- to 300-mL capacity.

2. Measure and record the temperature of the water being sampled.

3. Fix the sample by tipping the glass stopper away from the filled BOD bottle. Using a measuring pipet, add 1 mL manganous sulfate ($MnSO_4$) solution (step 1). Place the tip of the pipet below the surface of the water to allow the heavy solution to flow in without coming into contact with air.

4. In the same way, add 1 mL alkali-iodide-azide solution (step 2).

5. Carefully replace the glass stopper so no air is trapped below it.

6. Mix by inverting the bottle several times for 3 min.

7. Allow the resulting precipitate to settle to at least one-half the bottle height.

8. Invert the bottle several times more, then set aside until the precipitate has settled at least halfway in the bottle.

9. Tip the stopper again. With a measuring pipet, add 1 mL concentrated sulfuric acid (step 3).

10. Replace the stopper carefully to prevent air from entering the bottle. Rinse the outside of the stoppered bottle with tap water. Then mix by inverting several times until the precipitate completely dissolves and the brown or yellow is distributed uniformly.

11. Analyze the sample by transferring the 201 mL of sample to a wide-mouth 500-mL Erlenmeyer flask.

12. Fill a buret with sodium thiosulfate titrant (step 6), and record the liquid level by reading the bottom of the meniscus. Guard against a leaky stopcock, which results in the loss of titrant.

13. Gradually add small portions of sodium thiosulfate titrant (step 6) from the buret while constantly swirling the liquid in the flask, until the sample changes to a pale yellow or straw color.

14. With a measuring pipet, add to the flask 1 or 2 mL starch indicator solution (step 4), which will cause the solution to turn blue.

15. Continue adding sodium thiosulfate titrant (step 6) drop by drop until the blue disappears. Ignore any reappearance of the blue on standing.

16. Record the new buret level by reading at the bottom of the meniscus.

17. Calculate the volume of titrant used by subtracting the buret reading in (step 12) from the buret reading in (step 16). The result is the DO concentration in mg/L.

This page intentionally blank

Chapter **13**

Fluoride

PURPOSE OF TEST

Many water supplies add fluoride to reduce the incidence of dental cavities. A concentration of approximately 1 mg/L is an effective fluoride level without harmful health effects. Teeth may become mottled when fluoride concentration exceeds 2 mg/L. Severe dental fluorosis, causing teeth to become stained, pitted, and brittle, may occur when the concentration exceeds 4 mg/L. In rare instances the naturally occurring concentration may exceed 10 mg/L, in which case the source water should be de-fluoridated. Accurate determination of fluoride has increased in importance as the practice of fluoridating water supplies has grown. An optimal fluoride concentration is essential to maintain effective and safe levels of fluoridation.

LIST OF SIMPLIFIED METHODS

Refer to *Standard Methods for Examination of Water and Wastewater*, Section 4500 F. D. SPADNS Method. This method may require acid distillation of the fluoride sample. Refer to the Complexone (AutoAnalyzer) method, Section 4500–F.E and the Ion-Selective Electrode Method 4500–F.C. For more information, see the US Environmental Protection Agency's *Methods for Chemical Analysis of Water and Wastes*.

SIMPLIFIED PROCEDURE

Ion–Selective Electrode Method

Fluoride is determined potentiometrically with a fluoride electrode in conjunction with a standard single-junction sleeve-type (not fiber-tip) reference electrode and a pH meter with an expanded millivolt scale.

Warnings/Cautions.

Extremes of pH interfere. The pH of the sample should be between 5 and 9. Adding a fluoride 5.3–5.5 buffer (see Reagents) eliminates this problem. This buffer also contains a storing chelating agent that preferentially complexes aluminum, silicon, and iron.

These polyvalent cations (Si^{+4}, Fe^{+3}, and Al^{+3}) interfere with the electrode by forming complexes with fluoride. The degree of interference depends on the concentration of the complexing cations, the concentration of fluoride, and the pH of the sample. Using this buffer eliminates these problems.

The temperature of samples and standards must be within $1°C$. Place a thin piece of plastic or other insulating material between the magnetic stirrer and the sample in the beaker to prevent the stirrer from raising the sample's temperature.

Between samples, rinse the electrodes with distilled water and carefully blot dry. Blot the electrodes very gently to avoid damaging the sensing surface.

Collect samples in polyethylene bottles. Preservatives are not necessary. Holding times should not exceed 28 days.

Apparatus.

- a pH meter with expanded millivolt scale

- a fluoride ion activity electrode

- a sleeve-type reference electrode (not fiber-tip)

NOTE: **Follow the manufacturer's instructions for the care and conditioning of electrodes.**

- a magnetic stirrer and several TFE-coated stir bars

- a 150-mL beaker

- a 1-L beaker

- a timer

- a volumetric flask

- a polyethylene bottle

Reagents.

Stock fluoride solution. Dissolve 0.2210 g anhydrous sodium fluoride (NaF) in distilled water and dilute to 1 L in a volumetric flask. Store in a polyethylene bottle. 1.0 mL = 0.1 mg F.

Standard fluoride solution. Using a 100-mL volumetric pipet, dilute 100.0 mL of the stock fluoride solution to 1 L in a volumetric flask with distilled water. Store in a polyethylene bottle. 1.0 mL = 0.01 mg F^-.

Sodium hydroxide, 5N. Slowly dissolve 200 g sodium hydroxide (NaOH) in distilled water. Cool and dilute to 1 L. Work under fume hood, with beaker surrounded by running water or in ice water.

Fluoride buffer pH 5.3–5.5. To approximately 500 mL distilled water in a 1-L beaker add 57 mL glacial acetic acid, 58 g sodium chloride, and 4 g 1,2 cyclohexylene-diaminetetraacetic acid (CDTA).

NOTE: **Work under laboratory safety hood to avoid fumes from CDTA. Care should be taken to avoid contact with the skin and breathing the dust.**

Stir to dissolve and cool to room temperature. Place the beaker in a cool water bath and slowly add the 5*N* sodium hydroxide solution to the solution in the beaker while stirring until the pH is between 5.3 and 5.5. About 150 mL will be required. Transfer the solution to a 1-L volumetric flask and dilute to the mark with distilled water.

NOTE: It would be clearer to the analyst if the procedure said to add the 3 components to the solution and place on a stir plate. Insert a calibrated pH electrode into the solution. Add 125 mL of 5N NaOH to the solution. Slowly add 5N NaOH to the solution until the pH is between 5.3 and 5.5. It is also helpful to place the beaker of solution in a cold water bath while adding the NaOH because it tends to generate heat.

All reagents are available commercially. Appendix A provides a list of manufacturers and suppliers.

Procedure.

1. Place 50 mL of the sample or standard solution and 50 mL of the buffer in a 150-mL beaker.

2. Add the stirring bar and place on the stirrer.

3. Immerse the electrodes in the solution and start the stirrer at medium speed. The volume in the beaker should allow the stirrer free movement. Let the electrodes remain in the solution at least 3 min or until the reading stabilizes. Rinse and blot dry between readings.

4. At concentrations less than 0.5 mg/L F$^-$, the solution may require 5 min to reach a stable meter reading; higher concentrations may stabilize more quickly.

5. Record the potential measurement for each standard and sample and convert to fluoride ion concentration using the standard curve as described in the manufacturer's calibration instructions. Check the reading of a standard frequently.

NOTE: **Direct reading instruments are available. Follow the manufacturer's directions.**

Already diluted commercial standards are available (see appendix A). It may be necessary to verify their concentration by comparing them to standards prepared in the laboratory. For more information on preparing a standard curve, see the Standard Calibration Curves section in chapter 1.

Calibration

Prepare a series of standards using the fluoride standard solution (step 2 in the reagents section) to cover the range of expected sample concentration by diluting volumes to 50 mL. Use the series provided in Table 13-1. A minimum of five standards should be used to prepare the curve. When using a previously prepared curve, it should be validated with one standard at the lower end and one standard at the higher end of the curve. These standards should be within 10% of their true value. Use of a standard from an external source is recommended to verify the curve.

Calibrate the electrometer or pH meter. Proceed as described in step 1 of the procedure section. Using semilogarithmic graph paper, plot the concentration of fluoride in mg/L on the log axis. Then plot the electrode potential developed in the

Table 13-1   Dilutions to prepare fluoride standards

mL of Standard Dilution to 50 mL	Concentration by mg/L
0.00	0.00
1.00	0.20
2.00	0.40
3.00	0.60
4.00	0.80
5.00	1.00
6.00	1.20
8.00	1.60
10.00	2.00

standard solutions on the linear axis, starting with the lowest concentration at the bottom of the scale.

Calibrate a selective ion meter. Follow the manufacturer's directions for operating the instrument.

Chapter **14**

Hardness

PURPOSE OF TEST

Calcium and magnesium in water cause hardness. Total hardness is defined as the sum of calcium and magnesium concentrations, both expressed as calcium carbonate in mg/L. The ethylenediaminetetraacetic acid (EDTA) titrimetric method can be used to determine total hardness or calcium hardness. This method is designed for the routine determination of hardness in drinking water.

LIST OF SIMPLIFIED METHODS

Refer to *Standard Methods for the Examination of Water and Wastewater*, Section 2340 C. EDTA Titrimetric Method.

SIMPLIFIED PROCEDURE

Hardness, Titration Method

Warnings/Cautions.

Water should be free of color and turbidity that might obscure or affect the indicator response. Fortunately, substances that cause errors in this titration are seldom present in drinking water supplies.

Barium, strontium, cadmium, lead, zinc, and manganese interfere by causing fading or indistinct end points. Limited amounts of copper, iron, cobalt, nickel, and aluminum can also affect hardness results. Consult *Standard Methods* for the concentrations of these substances that can be tolerated in a sample.

Titration should be performed on a sample at room temperature. The reaction is slow at low temperatures, and the indicator decomposes at high temperatures.

Complete the titration within 5 min of the time the buffer is added to prevent the calcium carbonate from precipitating. Both buffer and solid indicator mixtures are subject to deterioration and should be kept tightly stoppered when not in use.

An indistinct color or green off-color end point means a fresh indicator is needed.

Apparatus.

- a 25-mL buret and support

- a 50-mL graduated cylinder or volumetric pipets for measuring

- two or more 250-mL flasks or 150-mL evaporating dishes

- a dropping pipet or medicine dropper for dispensing buffer

- a spatula for dispensing dry indicator mixture

Reagents.

The following solutions and solid mixtures are commercially available or may be prepared as follows.

Buffer solution.

1. On an analytical balance, carefully weigh 1.179 g dry disodium ethylene–diaminetetraacetate dihydrate (abbreviated EDTA or Na_2EDTA) of reagent-grade quality and 0.780 g magnesium sulfate ($MgSO_4 \cdot 7H_2O$). Carefully transfer the two chemicals to a 100-mL beaker and dissolve in 50-mL distilled water. An alternative odorless buffer with a longer shelf life is available commercially.

2. Weigh 16.9 g ammonium chloride (NH_4Cl) and place in a 400-mL beaker with a 250-mL graduated cylinder, measure 143 mL concentrated ammonium hydroxide (NH_4OH), and add to the 400-mL beaker. Dissolve ammonium chloride in the concentrated ammonium hydroxide.

3. Pour EDTA and magnesium sulfate solution into ammonium chloride and ammonium hydroxide solution while stirring.

4. Pour the combined solution back into the 250-mL graduated cylinder and dilute with distilled water to the 250 mL mark. Mix the resulting solution thoroughly by pouring back and forth between the 400-mL beaker and the 250-mL graduated cylinder. Store in a tightly stoppered bottle for no longer than 1 month. This solution is available commercially (see appendix A).

Eriochrome Black T indicator for total hardness procedure. Weigh 0.5 g Eriochrome Black T dye and 100 g sodium chloride (NaCl). Place both chemicals in a mortar and grind together with a pestle until the dark dye is uniformly distributed throughout the white salt. Store in a tightly stoppered bottle. During titration, the color changes from red to purple to bluish purple to a pure blue with no trace of red or purple tint at the end point.

Eriochrome Blue Black R indicator for calcium hardness procedure. Weigh 0.2 g Eriochrome Blue Black R dye and 100 g sodium chloride. Place both chemicals in a mortar and grind together with a pestle until the dark dye is uniformly distributed throughout the white salt. Store in a tightly stoppered bottle. During titration, the color changes from red to purple to bluish purple to a pure blue at the end point.

EDTA titrant (0.01M).

1. On an analytical balance, carefully weigh 3.723 g reagent-grade dry disodium ethylenediaminetetraacetate dihydrate (abbreviated EDTA or Na_2EDTA) of reagent-grade quality. Carefully transfer the chemical to a 250-mL beaker and dissolve in 150 mL distilled water.

2. Carefully transfer the solution to a 1-L volumetric flask, rinsing the beaker with three 100-mL portions of distilled water. Add rinsings to flask. Dilute to the 1 L mark with distilled water. Stopper and mix thoroughly. This solution is available commercially (see appendix A).

Standard calcium solution.

1. On an analytical balance, carefully weigh 1.000 g anhydrous calcium carbonate ($CaCO_3$) powder. Carefully transfer the chemical to a 500-mL flask. (This is available commercially.)

2. Place a glass funnel in the flask neck and add hydrochloric acid (HCl) (1+1), drop by drop, until all calcium carbonate has dissolved.

3. Add 200 mL distilled water and boil for a few minutes. Cool and add a few drops of methyl red indicator. Adjust to the intermediate orange color by adding ammonium hydroxide (3N) or HCl (1+1), as required.

4. Carefully transfer solution to a 1-L volumetric flask. Rinse the flask with three 100-mL portions of distilled water. Add rinsings to flask. Dilute to the 1-L mark with distilled water (1 mL = 1 mg $CaCO_3$).

Hydrochloric acid (1+1). Cautiously add and slowly mix 100 mL concentrated hydrochloric acid (HCl) to 100 mL distilled water.

Ammonium hydroxide (3N). Under the hood, dilute 50 mL concentrated ammonium hydroxide (NH_4OH) to 250 mL with distilled water.

Sodium hydroxide (1N). Dissolve 100 g sodium hydroxide (NaOH) in 100 mL distilled water. Dilute to 250 mL in distilled water.

Methyl red indicator. Dissolve 0.1 g methyl red sodium salt in distilled water and dilute to 100 mL.

Standardizing EDTA titrant procedure.

1. Fill the buret with EDTA titrant. Record the liquid level in the buret by reading at the bottom of the meniscus.

2. Place 10 mL standard calcium solution in a 250-mL flask. Add 40 mL distilled water to the flask.

3. Also prepare a color comparison blank by placing 50 mL distilled water in a 250-mL flask.

4. Add 1–2 mL buffer solution to each flask. Complete the titration within 5 min measured from the time of buffer addition.

5. Add approximately 0.2 g Eriochrome Black T indicator to each flask and mix.

6. To the color comparison blank, add EDTA titrant from the buret drop by drop, with continuous stirring, until the last reddish tinge disappears. At the end point, the solution is pure blue. Record the new buret level by

reading at the bottom of the meniscus. Sometimes no EDTA will be needed to turn the solution blue; at other times as many as 3 drops will be required.

7. Next, titrate the standard calcium solution by adding EDTA slowly, with continuous stirring, until the last reddish tinge disappears. Add the last few drops at 3- to 5-sec intervals. At the end point, the solution is the same blue as the color comparison blank.

8. Record the new buret level by reading at the bottom of the meniscus.

9. Calculate the net volume of titrant used by subtracting the initial buret reading (step 1) from the final reading (step 8).

10. Calculate the blank correction by subtracting the buret reading in step 1 from step 6.

11. Calculate the net volume of titrant used for the standard calcium solution alone by subtracting the result found in step 10 from the result found in step 9.

12. Calculate mg calcium carbonate equivalent to 1 mL EDTA as follows:

    mg $CaCO_3$ equivalent to 1 mL EDTA = 10/net mL EDTA used to titrate 10 mL standard calcium solution

13. Standardize EDTA monthly.

Total hardness procedure.

1. Fill the buret with EDTA. Record liquid level in the buret by reading at the bottom of the meniscus.

2. Measure the sample volume (Table 14-1) for the indicated hardness ranges.

    If a sample of only 25 mL is needed, add 25 mL distilled water to bring the total volume to 50 mL. If a sample of only 10 mL is needed, add 40 mL distilled water. Measure the additional distilled water with a graduated cylinder. Place the sample (and distilled water, if needed) in a 250-mL flask.

3. Prepare a color comparison blank by placing 50 mL distilled water, measured with a graduated cylinder, in a similar 250-mL flask.

4. Add 1–2 mL buffer solution to the color comparison blank and the sample and mix. Complete the titration within 5 min measured from time of buffer addition.

Table 14-1   Sample volumes for various calcium carbonate hardness ranges

Sample Volume mL	Hardness Range mg/L as $CaCO_3$
50	0–300
25	301–600
10	601–1,500

5. Add approximately 0.2 g Eriochrome Black T indicator to the color comparison blank and the sample and mix.

6. To the color comparison blank, add EDTA titrant drop by drop until the pink changes to pure blue. Sometimes no EDTA will be needed; at other times as many as 3 drops will be required. Record the new buret level by reading at the bottom of the meniscus.

7. If the sample turns pink or purple, add EDTA titrant from the buret. Stir the flask constantly. Continue adding titrant until the red turns to a purple tinge. Stop adding titrant for 10 sec but continue stirring.

8. Resume adding the EDTA titrant drop by drop until the purple turns the same pure blue as the color comparison blank. Stir the flask constantly. The change from purple to pure blue occurs within a span of 1 to 4 drops. If you have difficulty recognizing the change when adding 1 drop at a time, add the titrant 2 drops at a time near the end point. This intensifies the color change, and the slight loss in accuracy is not significant.

9. Record the new buret level by reading at the bottom of the meniscus.

10. Calculate the gross volume of titrant used by subtracting the initial buret reading (step 1) from the last reading (step 9).

11. Calculate the blank correction by subtracting the buret reading in step 1 from the buret reading in step 6.

12. Calculate the net volume of titrant used for the sample alone by subtracting the result found in step 11 from the result found in step 10.

13. Calculate total hardness as follows:

$$\text{Hardness as mg } CaCO_3/L = \frac{A \times B \times 1,000}{\text{mL sample}}$$

Where:

A = mL titrant used for sample (see step 12)

B = mg $CaCO_3$ equivalent to 1 mL EDTA titrant (see step 12)

Calcium hardness procedure.

1. Fill the buret with EDTA titrant. Record the liquid level in the buret by reading at the bottom of the meniscus.

2. Prepare the sample volume or dilution for the indicated hardness range (see previous step 2). Place the sample and distilled water, if necessary, in a 250-mL flask.

3. Prepare a color comparison blank by placing 50 mL distilled water in a 250-mL flask.

4. Add 2 mL sodium hydroxide solution (or enough to produce a pH of 12 to 13) to each flask.

5. Add approximately 0.2 g Eriochrome Blue Black R indicator to each flask.

6. To the color comparison blank, add EDTA titrant drop by drop until the pink changes to pure blue. Sometimes no EDTA will be needed; at other times, as many as 3 drops will be required. Record the new buret level by reading at the bottom of the meniscus.

7. If the sample turns pink or purple, add EDTA titrant from the buret. Stir the flask constantly. Continue adding the titrant until the red turns to a purple tinge. Stop adding titrant for 10 sec but continue stirring.

8. Resume adding the EDTA titrant drop by drop until the purple turns the same pure blue as the color comparison blank. Stir the flask constantly. The change from purple to pure blue will occur within a span of 1 to 4 drops. If you have difficulty recognizing the change when adding 1 drop at a time, add the titrant 2 drops at a time near the end point. This will intensify the color change, and the slight loss in accuracy is not significant.

9. Record the new buret level by reading at the bottom of the meniscus.

10. Calculate the gross volume of titrant used by subtracting the initial buret reading (step 1) from the last reading (step 9).

11. Calculate the blank correction by subtracting the buret reading in step 1 from the buret reading in step 6.

12. Calculate the net volume of titrant used for the sample alone by subtracting the result found in step 11 from the result found in step 10.

13. Calculate calcium hardness as follows:

$$\text{Hardness as mg CaCO}_3/\text{L} = \frac{A \times B \times 1,000}{\text{mL sample}}$$

Where:

A = mL titrant used for sample

B = mg $CaCO_3$ equivalent to 1.00 mL EDTA titrant

Magnesium hardness by calculation procedure.

Calculate magnesium hardness as

mg $CaCO_3/L$ = Total Hardness (as mg $CaCO_3/L$) – Calcium Hardness (as mg $CaCO_3/L$).

Chapter **15**

Iron

PURPOSE OF TEST

Iron causes complaints about the aesthetic water quality of many groundwaters. These problems become more noticeable when the iron concentration exceeds 0.5 mg/L. At concentrations below 1 to 1.5 mg/L, iron-related problems may be partially controlled by adding ortho- or polyphosphate at a dosage level of 2 to 4 mg/L per mg/L of iron. Under some conditions, phosphate will sequester or chelate iron. The phosphate-chelating bond with iron breaks down with time and in water heaters. Problems with higher iron concentrations are addressed by iron removal systems that oxidize iron through chemical addition or aeration followed by filtration.

The measurement of iron in drinking water is mainly concerned with determining the level of ferrous iron or unoxidized (dissolved) iron still remaining in solution. The goal of the water system operator is to mitigate the aesthetic problems while iron is still in solution. Options for controlling ferric or oxidized iron (precipitated or undissolved) are limited.

LIST OF SIMPLIFIED METHODS

Atomic absorption (AA) spectrophotometric methods are not simplified methods for determining the concentration of iron in drinking water. However, they produce accurate analytical results. Atomic absorption methods include *Standard Methods* Sections 3111 B. and 3113 B., and US Environmental Protection Agency (USEPA) Method 200.9.

Inductively coupled plasma (ICP) methods are not simplified methods for determining the concentration of iron in drinking water. They produce accurate analytical results. Common ICP methods include USEPA Methods 200.7 and 200.8, and *Standard Methods* 3120 B.

Test kit methods are available from several manufacturers. Their suitability depends on the intended use of the results, as their accuracy and precision are highly variable. Despite these limitations, test kits offer considerable savings in labor and analytical and equipment costs.

The phenanthroline method produces accurate results under most circumstances and offers a saving on equipment cost at the expense of manpower. Refer to *Standard Methods*—Method 3500–Fe B.

SIMPLIFIED TEST PROCEDURES

The procedure for determining dissolved iron concentration using a photometer or spectrophotometer has been modified.

Warnings/Cautions.

The presence of orthophosphate, polyphosphates, free or total chlorine, or chlorine dioxide interferes with the analysis. Also, typically not present in drinking water, high concentrations of nitrite, cyanide, and certain metals interfere with the analysis of iron. Similarly, naturally occurring color, turbidity, or organic matter may cause errors. Additional steps may be needed to compensate for turbidity or color.

This modification to *Standard Methods* assumes that concentrations of interferences are at minimal or controllable levels and is intended to produce results that are satisfactory for process control work.

Caution: Exercise extreme care when handling acids during sample processing.

Apparatus.

- a spectrophotometer or photometer equipped to measure light at or near 510 nm

- 125-mL or larger funnels

- filter paper, minimally acceptable general purpose; ideal is 0.45 μm pore size

- 125-mL Erlenmeyer flasks or beakers

- a 100-mL graduated cylinder

- one each 100-, 500-, and 1,000-mL volumetric flasks

- pipets

- a hot plate and thermometer (optional)

- a magnetic stirrer (optional)

- goggles, apron, and gloves for handling acids

Reagents.

Reagent-grade or distilled water.
Concentrated, glacial-grade acetic acid.
Ammonium acetate ($NH_4C_2H_3O_2$).
Phenanthroline monohydrate. Dissolve 100 mg phenanthroline monohydrate ($C_{12}H_8N_2 \cdot H_2O$) in 100 mL reagent-grade water by stirring and gentle heating. Temperature should approach 80° C and should not boil. Alternatively, add 2 drops of concentrated HCl to the water. Discard the solution if it darkens.

Hydrochloric acid (HCl)
Sulfuric acid (H_2SO_4)
Ferrous ammonium sulfate [$Fe(NH_4)_2(SO_4)_2 \cdot 6HO$]
Potassium permanganate ($KMnO_4$)

Procedure.

1. Prepare a working stock iron solution in an Erlenmeyer flask or a beaker. Add 20 mL concentrated sulfuric acid (H_2SO_4) to 50 mL reagent-grade water. Add and dissolve 1.404 g ferrous aluminum sulfate. While stirring, add $0.1N$ potassium permanganate ($KMnO_4$) drop by drop until a slightly pink color remains. Transfer solution to 1,000–mL volumetric flask. Using reagent-grade water, dilute to the mark (1 mL = 200 µg Fe).

2. Prepare a series of iron calibration standards. Add the volume of working iron standard as shown in Table 15-1 to the 1,000-mL volumetric flask, then dilute to the mark with reagent-grade water. The calibration curve should bracket the iron concentration being measured. The curve should contain a blank and at least three other points. Create a plot of mg/L iron versus absorbance.

3. Add 2 mL concentrated hydrochloric acid to 100 mL sample.

4. Transfer 50 mL acidified sample to a 100-mL flask or beaker.

5. For samples that contain 1 mg/L or less dissolved iron, add 20 mL phenanthroline solution and 10 mL ammonium acetate buffer solution to the 50-mL portion of acidified sample. For samples with more than 1 mg/L dissolved iron, add 20 mL phenanthroline for each additional 1 mg/L of iron. Ensure sample is well mixed or stirred. For samples with more than 2 mg/L of iron, dilute the sample with reagent-grade water before adding the phenanthroline or ammonia acetate buffer.

6. Dilute to 100 mL using reagent-grade water.

7. Measure color using the photometer or spectrophotometer with a filter or setting at 510 nm.

Table 15-1   Preparing iron calibration standards

Working Stock Iron Solution mL	Calibration Standard mg/L as Fe
0	0 (Blank)
1	0.2
2	0.4
4	0.8
5	1.0
8	1.6
10	2.0

This page intentionally blank

Chapter **16**

Jar Tests

PURPOSE OF TEST

Jar tests show the nature and extent of the chemical treatment that will optimize the quality of water that leaves the plant. Many chemicals added to a water supply can be evaluated on a laboratory scale using a jar test. Among the most important chemicals are coagulants, coagulant aids, alkaline compounds, softening chemicals, and powdered activated carbon for taste and odor removal and adsorption of organics. When coagulants are added, the jar test may be called a coagulation or flocculation test.

Surface waters generally contain suspended particles that cause turbidity. These particles vary in size and amount. Turbidity can be removed by adding a coagulant such as alum, followed briefly by rapid mixing, then a longer period of slow mixing (flocculation), and finally a period of settling. The laboratory coagulation test that uses jar testing equipment is an attempt to imitate the mixing, flocculation, and settling conditions in the full-scale treatment plant.

Select a series of doses so the first jar represents undertreatment and the last jar represents overtreatment. This approach indicates the correct coagulant dosage for the plant when varying amounts of turbidity, color, or other factors dictate a change in the coagulant dose. When this information is fine-tuned and implemented, it helps the plant produce a clear, colorless water.

Jar tests can also be used to compare the performance of various types of coagulants and coagulant aids before they are used in the full-scale treatment plant. Separate tests for each coagulant can be performed on the same water. Floc formation, turbidity removal, color removal, pH, and cost per thousand gallons can then be compared for the optimal dose of each coagulant.

LIST OF SIMPLIFIED METHODS

Jar tests are conducted using commercially available stirring machines that have three to six paddles. Tests can be conducted to determine correct dosages and treatment regimes for coagulation, lime or soda ash softening, and activated carbon addition. After chemical treatment, samples are collected and analyzed for turbidity, pH, hardness, odor, organic compounds, or any other parameter of interest. See other chapters of this manual for descriptions of these tests.

For further information, see *Operational Control of Coagulation and Filtration Processes* and the Water Supply Operations series video on *Jar Testing*.

SIMPLIFIED PROCEDURES

General Jar Test Information

Warnings/Cautions.

Even the smallest detail may have an important influence on the result of a jar test. Therefore, all samples in a series of tests should be handled as similarly as possible.

The purpose of the test determines the necessary experimental conditions, e.g., stirring speed, the length of the flash mix, flocculation, and settling intervals. The best results are achieved when jar test conditions closely match those used in the full-scale plant for each of these controllable parameters.

Temperature plays an important role in coagulation, so source water samples should be collected only after all other preparations have been made. Source water may be stored for longer periods of time if the water can be kept at the same temperature as the water being treated, i.e., submerged in a water bath.

Apparatus.

- a stirring machine with three to six paddles that can operate at variable speeds (0 to 270 rpm), shown in Figure 1-39

- a floc illuminator for observing floc formation (optional)

- special, square, 1- or 2-L jars with side sample ports, sometimes called gator jars (conventional beakers or circular jars with baffles can be used, but sample collection is more difficult)

- a large plastic carboy for collecting source water

- a thermometer

- top-loading balance

- 1-L volumetric flasks

- disposable plastic syringes, such as B-D Plastipak syringes or equivalent, for dosing samples rapidly with coagulants and other solutions (most common sizes are 3-, 5-, 10-, and 20-mL volumes)

- 100-mL pipets for obtaining samples if gator jars with side ports are not used

- sample containers, such as small beakers, to collect treated jar test samples

- analytical equipment required for tests such as turbidity and hardness

Reagents.

Dosing solutions. Suspensions or solutions prepared from the stock materials actually used in the plant.

Distilled water. Boil distilled water for preparing lime suspensions for 15 min to expel carbon dioxide.

Procedure.

Always select a series of doses for the chemical being tested so the first jar represents undertreatment and the last represents overtreatment. When a proper series is set up, the succession of jars will show poor, fair, good, and excellent results at the end of the test. In the case of coagulation, overtreatment may also yield poor results. Repeat the jar test once or twice to find the proper series of doses for the desired results. The objective is to determine the lowest (and therefore the least expensive) dose that will produce the desired water quality.

Rapid mix and flocculation mixing intensities and stirring times should imitate those used in the full-scale plant, when possible. Because of differences in energy transfer to the water in the jar by the stirrer and in the water in the treatment plant by the rapid mix and flocculation impellers, detention time may differ between the jar and treatment plant basins. Scale calculations may be required. (See GT values.)

Plant operating conditions may also dictate a change in the laboratory stirring times. For example, the flocculation period in a plant that operates at 5 mgd during winter may be 40 min, but a summer pumping rate of 10 mgd reduces the flocculation period to 20 min. Make similar downward adjustments when a plant has parallel flocculation basins, but one is taken out of service, thereby reducing the effective flocculation period.

Coagulation Jar Test

Warnings/Cautions.

Refer to the general jar test information discussed earlier in this chapter.

Many coagulants are irritants to skin, and contact with concentrated solutions should be avoided. Wash thoroughly with water if contact occurs. Read Material Safety Data Sheets for specific information.

Dilute coagulant solutions have a very limited shelf life. Follow the instructions listed below for each coagulant type.

Apparatus.

See the General Jar Test Information section earlier in this chapter.

- 1-L and 250-mL beakers

- a 500-mL volumetric flask

- a hand-held electric blender

Reagents.

Many chemicals and other materials are used to coagulate turbid or colored waters. Primary coagulants are generally inorganic and used alone or in conjunction with coagulant aids.

Alkalinity, which is sometimes required to produce the desired hydrous oxide floc, can be added in the form of calcium hydroxide (hydrated or slaked lime), calcium oxide (quicklime or unslaked lime), or soda ash.

Clay suspensions that consist of clay, kaolin, bentonite, or stone dust may be added to improve the coagulation of low-turbidity water from which it is difficult to form floc nuclei.

Inorganic coagulants such as aluminum sulfate (alum), ferric chloride, ferric sulfate, and polyaluminum chloride.

1. These coagulants should be prepared daily. If possible, obtain concentrated coagulant from the supply actually used in the plant.

2. Weigh 10 g of the material. If the coagulant is a liquid, this is most easily done by drawing it into a plastic syringe and weighing it on the balance, after first weighing the syringe empty or taring the balance.

3. Dissolve in 500 mL distilled water in a 1-L volumetric flask. Dilute to 1 L with distilled water. Mix thoroughly before each use. Each 1 mL of the resulting solution represents 5 mg/L as coagulant product in a 2-L water sample.

NOTE: Only a certain portion of the coagulant product is actually the working coagulant chemical. For example, alum may be 48.9 percent $Al_2(SO_4)_3$, thus 1 mL of the resulting solution prepared above would represent 2.4 mg/L $Al_2(SO_4)_3$, in a 2-L water sample. Therefore, be sure to account for the percentage of active ingredient when calculating dosages for inorganic coagulants.

Liquid polymer stock solution. Polymers are organic materials that can be used as coagulant aids. Polymers are classified as cationic (positively charged), anionic (negatively charged), or nonionic (no charge).

Prepare a 1 percent weight to volume solution.

1. Use a volumetric flask to measure 500 mL distilled water. Place the water into a 1-L beaker, and then use a syringe to withdraw 5 mL water. This leaves 495 mL water in the beaker.

2. Place a 5-mL plastic syringe on a top-loading balance and record the weight or tare the balance.

3. Shake the neat polymer until thoroughly mixed. Fill the syringe with 5 g neat polymer (about 4.8–4.9 mL). Weigh the syringe on the balance. Add or remove polymer from the syringe until the correct weight is reached.

4. Place the hand-held electric blender into the dilution water in the 1 L beaker. Start the blender, place the syringe tip below the water surface in an area of high turbulence and slowly empty the polymer from the syringe. Be careful not to inject the polymer directly into the beaker walls or blender surfaces.

5. Continue to mix for 30–60 sec. Let this stock solution age at least 15 min before preparing the 0.1 percent working solution.

6. Prepare this solution every 48 hr.

Prepare a 0.1 percent working solution.

1. Place 180 mL distilled water in a 250-mL beaker. Fill a 20-mL syringe with 20 mL 1 percent stock polymer solution. Be sure there are no air bubbles present.

2. Place the hand-held blender into the dilution water in the beaker. Start the blender, place the syringe below the water surface in an area of high turbulence, and then slowly empty the polymer from the syringe. Be careful not to inject the polymer directly into the beaker walls or blender surfaces.

3. Prepare fresh working solutions daily, preferably less than 2 hr before use.

4. Each 1 mL of the resulting working solution represents 0.5 mg/L in a 2-L water sample.

Emulsion polymer stock solution.

Prepare a 1 percent weight to volume stock solution.

1. Follow the same steps as for a liquid polymer. Be extra sure to thoroughly mix the neat polymer solution and work rapidly through steps 2 and 3 in succession.

2. Shake the neat polymer until thoroughly mixed. Fill the syringe with 5 g neat polymer (about 4.8–4.9 mL). Weigh the syringe on the balance. Add or remove the polymer from the syringe until the correct weight is obtained.

3. Place the hand-held electric blender into the dilution water in the 1-L beaker. Start the blender, place the syringe tip below the water surface in an area of high turbulence, and then slowly empty the polymer from the syringe. Be careful not to inject the polymer directly onto the beaker walls or blender surfaces.

Prepare a 0.05 percent working solution.
Follow the same steps as for a liquid polymer, except use 190 mL dilution water and add to it 10.0 mL stock polymer solution. Each 1 mL of the resulting working solution represents 0.25 mg/L in a 2-L water sample.

Solid polymer stock solution.

Prepare a 0.5 percent weight to volume stock solution.

1. Use a volumetric flask to measure 500 mL distilled water. Place the water in a 1-L beaker.

2. Use a weighing boat and a top-loading balance to weigh out 2.5 g polymer. Work quickly, as the solid may draw water from the air.

3. Immediately place the hand-held blender into the water in the 1-L beaker and begin to stir. Slowly add the solid polymer to the mixing water in an area of high turbulence. Do not let the solid particles contact the beaker walls or blender surfaces as they are being sprinkled into the dilution

water. Pause to let the solution mix after each incremental addition of polymer solid.

4. Continue to mix for 60–90 sec or until the solution appears homogeneous. Let the solution age for at least 15 min before further dilution.

5. Prepare every 48 hr.

Prepare a 0.01 percent working solution:
Follow the same steps as for a liquid polymer, except use 196 mL dilution water and add to it 4.0 mL stock polymer solution. Each 1 mL of the resulting working solution represents 0.05 mg/L in a 2-L water sample.

Additional reagents.

Lime or Soda Ash Solutions.

1. Distilled water used to prepare lime suspensions should be boiled for 15 min to expel the carbon dioxide. Cool to room temperature shortly before adding lime.

2. Weigh 10 g of solid material on a top-loading balance. Suspend in 1 L distilled water in a volumetric flask. Mix thoroughly immediately before each use.

3. Each 1 mL dosing solution represents 5 mg/L in a 2-L water sample.

4. Prepare daily.

Clay suspensions such as clay, kaolin, bentonite, or stone dust.

1. Weigh out 10 g of the material.

2. Suspend in 1 L distilled water.

3. Each 1 mL of the resulting suspension represents 5 mg/L in a 2-L water sample.

4. Shake vigorously immediately before use.

5. Prepare monthly.

Procedure.

1. Clean the gator jars using detergent and a scrub brush. Rinse thoroughly with tap water.

2. Clean the stirring machine paddles with a damp cloth.

3. Collect the water to be treated in a plastic carboy. Unless the water can be stored at ambient temperature, begin the jar test procedure promptly so the water temperature does not change significantly.

4. Mix the sample water thoroughly and fill all the gator jars half full. Remix the sample water. In reverse order, fill the gator jars to the fill line.

5. Place the paddles in the jars and tighten the thumb screws.

6. If using a clay suspension, add it to each jar with a wide-tip pipet, a plastic syringe, or a graduated cylinder.

7. Fill plastic syringes with the correct amounts of coagulant or coagulant aid solution to be added to each jar. Ensure there are no air bubbles, especially near the tip of the syringe. If bubbles are present, empty and refill the syringe. Fill one syringe for each jar.

8. Begin rapid mixing by setting the machine to the desired paddle speed (rpm value). In a 2-L square gator jar a mixing speed of 270 rpm corresponds to a mixing intensity of about 500 s^{-1}.

9. Inject the coagulant solutions into the jars below the surface of the water near the paddles. Inject all the jars at the same time for each type of coagulant solution added.

10. Add chemicals in the order they would be added in the plant and mix each one for the same time period mixing would occur in the plant.

11. If desired, use plastic syringes to add lime or soda ash to add alkalinity to waters deficient in natural alkalinity. See Table 16-1 for the amounts of lime or soda ash needed to neutralize various coagulants.

12. Lower the paddle speed to simulate the flocculation mixing process. If possible, mix the water at the same mixing gradient and for the same time period as occurs in the plant. Figure 16-1 shows the correlation between rpms in the gator jar and the mixing intensity within the jar.

13. Observe each jar for the appearance of "pinpoint" floc and note its time and order.

14. At the end of the flocculation period, stop the stirring machine, remove the paddles, and allow the floc to settle.

15. Observe the floc as it settles and describe the results. A hazy sample indicates poor coagulation. Properly coagulated water contains well-formed floc particles, and the liquid between the particles is clear.

16. Collect samples after 5, 10, 15, and 30 min or as desired. Collect samples by opening the clamp on the sample tube on the side of the gator jar, flushing 20 mL to waste, and collecting the necessary amount of sample into a sample cup. Do not clamp the tube shut between the flush and the sample collection, and make sure the tube is 100 percent open. If beakers are being used instead of gator jars, obtain the sample carefully with a 100-mL pipet.

Table 16-1   Chemical addition of alkalinity for jar testing

| Chemical | mg/L required to react with 1 mg/L of a given coagulant | | | |
	Total Alkalinity (as $CaCO_3$)	Soda Ash (as Na_2CO_3)	Lime [as $Ca(OH)_2$]	Lime (as CaO)
Alum	0.50	0.54	0.39	0.33
Ferric Sulfate	0.75	0.82	0.59	0.50
Ferric Chloride	0.92	1.0	0.72	0.61

NOTE: These dosages provide a starting point. Actual dosages are determined by the jar test.

17. Analyze the samples for turbidity, pH, and any other parameters of interest using the methods described in other sections of this manual.

Lime or Soda Ash Softening Jar Test

Warnings/Cautions.

Even the smallest detail may have an important influence on the result of a jar test. Therefore, all samples in a series of tests should be handled as similarly as possible.

The purpose of the test determines the experimental conditions, e.g., stirring speed, the length of the flash mix, flocculation, and settling intervals. The best

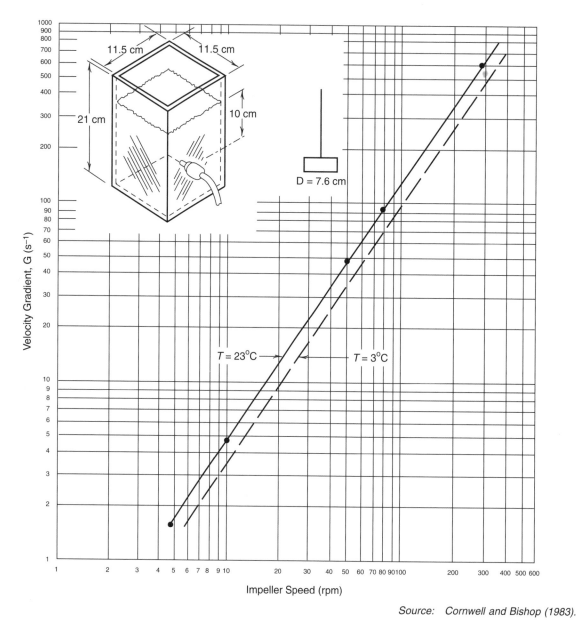

Source: Cornwell and Bishop (1983).

Figure 16-1   Laboratory G curve for flat paddle in the gator jar

results are achieved when conditions in the jar test closely match those used in the full-scale plant for each controllable parameter.

Temperature plays an important role in coagulation, so source water samples should be collected only after all other preparations have been made. Source water may be stored for longer periods if it can be kept at outside ambient temperatures.

Apparatus.

See the General Jar Test Information section earlier in this chapter.

- a stirring machine with three to six paddles that can operate at variable speeds (0–270 rpm)

- a floc illuminator for observing floc formation (optional)

- special, square, 1- or 2-L jars with side sample ports, sometimes called gator jars (conventional beakers can be used, but sample collection is more difficult)

- a large plastic carboy for collecting source water

- a thermometer

- a top-loading balance for weighing reagent chemicals

- 1-L volumetric flasks for preparing solution

- disposable plastic syringes for dosing samples rapidly with coagulants and other solutions (the most common sizes are 3-, 5-, 10-, and 20-mL volumes)

- 100-mL pipets for obtaining samples if gator jars with side ports are not used

- sample containers, such as small beakers, to collect treated jar test samples

- analytical equipment required for procedures such as turbidity and hardness

- filtration funnel and apparatus

- filter paper of medium retentiveness

Reagents.

Lime–dosing solution.

1. To prepare a lime suspension use distilled water that has been boiled for 15 min to expel the carbon dioxide. Cool to room temperature shortly before adding lime.

2. Weigh 10 g lime on a top-loading balance. Suspend in 1-L distilled water in a volumetric flask. Mix thoroughly immediately before each use.

3. Each 1 mL of dosing solution represents 5 mg/L in a 2-L water sample.

4. Prepare daily.

Procedure.

1. For the water to be treated, determine the hardness, phenolphthalein and total alkalinity, calcium, and free carbon dioxide as described in other sections of this manual.

2. Calculate the chemical dosages of lime or soda ash, or both, needed to soften the source water. Lime dosage for softening can be estimated using equation 16-1.

$$\text{Lime Feed mg/L} = \frac{(A + B + C + D)1.15}{\text{Purity of lime as a decimal}} \tag{16-1}$$

Where:

A = CO_2 in source water

$$(\text{mg/L as } CO_2) \times \frac{56}{44}$$

B = total alkalinity in source water

$$(\text{mg/L as } CaCO_3) \times \frac{56}{100}$$

C = hydroxide alkalinity in softener effluent

$$(\text{mg/L as } CaCO_3) \times \frac{56}{100}$$

D = magnesium in source water

$$(\text{mg/L as } Mg^{2+}) \times \frac{56}{24.3}$$

1.15 = excess lime dosage (15%)

NOTE: If using hydrated lime [Ca(OH)₂], substitute 74 for 56 in A, B, C, and D.

Soda ash dosage can be estimated using the following equation:

$$\begin{array}{l}\text{Soda Ash Feed} \\ \text{mg/L } Na_2CO_3\end{array} = \begin{array}{c}\text{(Noncarbonate Hardness)} \\ \text{as mg/L } CaCO_3\end{array} \times \frac{106}{100} \tag{16-2}$$

NOTE: These dosages provide a starting point. Actual dosages are determined by the jar test.

3. Follow steps 1–5 of the procedure for coagulation jar tests discussed earlier in this chapter.

4. Begin mixing by setting the paddle speed of the mixing machine to 30 rpm or a speed that duplicates the mixing intensity in the full-scale plant.

Figure 16-1 shows the correlation between rpms in the gator jar and the mixing intensity within the jar.

5. Fill plastic syringes with the correct amount of lime or soda ash to be added to each jar. Ensure there are no air bubbles, especially near the tip of the syringe. If air bubbles are present, empty and refill the syringe. For each dosing solution fill one syringe for each jar.

6. Inject the dosing solutions into the jars below the surface of the water near the paddles. Add chemicals in the same order they would be added in the plant.

7. Continue mixing for the same time period as in the plant.

8. Stop the stirring machine and allow the sample to settle until the liquid becomes fairly clear, usually about 10 to 15 min.

9. Collect samples by opening the clamp on the sample tube on the side of the gator jar, flushing 20 mL to waste, and collecting the necessary amount of sample. If beakers are used instead of gator jars, obtain the sample with a 100-mL pipet.

10. Warm the samples to 25°C.

11. Filter samples through filter paper to remove any solid precipitate.

12. Determine the hardness, phenolphthalein and total alkalinity, and pH as described in other chapters of this manual.

Powdered Activated Carbon Jar Test

Warnings/Cautions.

Even the smallest detail may have an important influence on the result of a jar test. Therefore, all samples in a series of tests should be handled as similarly as possible.

The purpose of the test determines experimental conditions, e.g., stirring speed, the length of the flash mix, flocculation, and settling intervals. The best results are achieved when conditions in the jar test closely match those in the full-scale plant for each controllable parameter.

Temperature plays an important role in coagulation, so source water samples should be collected only after all other preparations have been made. Source water can be stored for longer periods if it is kept at outside ambient temperatures.

Apparatus.

See the General Jar Test Information section earlier in this chapter.

- a stirring machine with three to six paddles that can operate at variable speeds (0–270 rpm)

- a floc illuminator to observe floc formation (optional)

- special, square, 1- or 2-L jars with side sample ports, sometimes called gator jars (conventional beakers can be used, but sample collection is more difficult)

- a large plastic carboy for collecting source water

- a thermometer

- a top-loading balance

- 1-L volumetric flasks for preparing solution

- disposable plastic syringes for dosing samples rapidly with coagulants and other solutions (most common sizes are 3-, 5-, 10-, and 20-mL volumes)

- 100-mL pipets for obtaining samples if gator jars with side ports are not used

- sample containers, such as small beakers, to collect treated jar test samples

- analytical equipment required for individual tests such as turbidity and hardness

- filtration funnel and apparatus

- glass wool or glass fiber filter paper that has been washed with odor-free water

Reagents.

Powdered activated carbon dosing solution.

1. Weigh 10 g powdered activated carbon on a top-loading balance. Dissolve in 1 L distilled water in a volumetric flask. Mix thoroughly immediately before each use.

2. Each 1 mL of the dosing solution represents 5 mg/L in a 2-L water sample.

Powdered activated carbon used in the plant may already be in slurry form. If this is the case, determine the concentration of the slurry by drying the slurry in an oven at 100° C. To do this, use a pipet to measure 10 mL well-mixed carbon slurry and place it into a small pre-weighed beaker. Place the beaker in the oven overnight. Allow to cool in a desiccator, and weigh the beaker. The weight of the dry carbon in milligrams divided by 0.01 gives the carbon concentration of the slurry in mg/L.

Carbon slurries can be stored for several months. Prepare or obtain a fresh solution every 6 months.

Procedure.

1. Clean and scrub beakers, stirring machine paddles, plastic carboy, and sample collection containers with a nonodorous detergent. Rinse thoroughly with odor-free water.

2. Follow steps 3 to 5 for coagulation jar tests discussed earlier in this chapter.

3. Begin mixing by setting the paddle speed of the mixing machine to 30 rpm or the speed that best duplicates the mixing intensity used in the full-scale plant. Figure 16-1 shows the correlation between rpm in the gator jar and mixing intensity within the jar.

4. Fill plastic syringes with the correct amount of carbon slurry to be added to each jar. Ensure there are no air bubbles, especially near the tip of the syringe. If air bubbles are present, empty and refill the syringe. One jar should have no carbon added so it can serve as a blank. Also fill syringes

for any other chemicals, such as coagulants, that are to be added during the jar test.

5. Inject the dosing solutions into the jars below the surface of the water near the paddles. Add chemicals in the same order they would be added in the plant.

6. Continue mixing for the same period as in the plant.

7. Stop the stirring machine and allow the samples to settle for about 15 min or for the desired time period.

8. Collect samples by opening the clamp on the sample tube on the side of the gator jar, flushing 20 mL to waste, and collecting the necessary amount of sample. If samples are being analyzed for organic parameters, collect samples directly into vials.

9. Filter a portion of each sample through a plug of glass wool or through glass fiber filters that have been previously washed with odor-free water. This will remove carbon from the samples to be analyzed for odor.

10. Determine the threshold odor of each treated sample and the blank as described elsewhere in this manual.

This page intentionally blank

Chapter **17**

Manganese

PURPOSE OF TEST

Manganese may cause complaints about the aesthetic quality of the drinking water. These problems become more noticeable when the manganese concentration exceeds 0.1 mg/L and are addressed by oxidation and filtration. The measurement of manganese in drinking water is mainly concerned with determining the level of unoxidized (dissolved) manganese in solution.

LIST OF SIMPLIFIED METHODS

Atomic absorption (AA) spectrometric methods are not simplified methods for determining the concentration of manganese in drinking water. However, they produce accurate analytical results. AA methods include *Standard Methods* Sections 3111 B. and 3113 B., and US Environmental Protection Agency (USEPA) Method 200.9.

Inductively coupled plasma (IPC) methods are not simplified methods for determining the concentration of manganese in drinking water, but they can produce accurate analytical results. Common ICP methods include USEPA Methods 200.7 and 200.8, and *Standard Methods* Section 3120 B.

Test kit methods are available from several manufacturers. Their suitability depends on the intended use of the results, as the accuracy and precision are highly variable. Despite these limitations, test kits offer considerable savings in labor and analytical and equipment costs.

The persulfate method can produce usable results and saves on equipment cost and labor expenses. Refer to *Standard Methods*, 3500–Mn B.

SIMPLIFIED PROCEDURE

The procedure has been modified for determining the dissolved manganese concentration using a spectrophotometer or photometer. The *Standard Methods* procedure relies on the use of mercuric sulfate to minimize the effects of high chloride levels. To avoid hazardous waste disposal problems, the simplified test procedure does not use mercuric sulfate.

Warnings/Cautions.

The presence of free or total chlorine or chlorine dioxide interferes with the analysis. Also, high concentrations of chloride interfere with the analysis of manganese. Similarly, naturally occurring color, turbidity, or organic matter may cause errors. Additional steps, not included in this modification, may be needed to compensate for turbidity or color.

This modification to the *Standard Methods* procedure assumes that concentrations of interferences are at minimal or controllable levels; it is intended to produce results satisfactory for process control work.

NOTE: **Exercise extreme caution when handling acids during sample processing.**

Apparatus.

- a spectrophotometer or photometer equipped to measure light at or near 525 nm

- 125-mL Erlenmeyer flasks or beakers

- a 100-mL graduated cylinder

- one each, 100-, 500-, and 1,000-mL volumetric flasks

- 1- and 2-L beakers

- pipets

- hot plate and thermometer

- magnetic stirrer (optional)

- goggles, apron, and gloves for handling acids

Reagents.

Reagent-grade or distilled water.
Concentrated, glacial-grade acetic acid (CH$_3$COOH).
Ammonium persulfate ([NH$_4$]$_2$S$_2$O$_8$).
Phenanthroline monohydrate (C$_{12}$H$_8$N$_2$•H$_2$O).
Nitric acid (HNO$_3$).
Sulfuric acid (H$_2$SO$_4$).
Potassium permanganate. Prepare a potassium permanganate (KMnO$_4$) stock solution in a 2-L beaker or 1-L volumetric flask. Add 3.2 g potassium permanganate and dilute to 1 L with reagent-grade or distilled water. Slowly heat until temperature approaches 90° C, but do not boil. Filter solution through fritted-glass filter crucible.
Sodium oxalate. Prepare a sodium oxalate (Na$_2$C$_2$O$_4$) solution to standardize the potassium permanganate stock solution. Record the weight of about 100 mg of sodium oxalate to the nearest 0.1 mg. Add sodium oxalate and 100 mL distilled water to a 1-L beaker. After dissolving the sodium oxalate, add 10 mL of sulfuric acid to the beaker and rapidly heat to a temperature of 90° to 95° C, but do not boil.
Sodium bisulfite (NaHSO$_3$).
Sodium nitrite (NaNO$_2$).
Hydrogen peroxide (H$_2$O$_2$).

Procedure.

1. While the sodium oxalate solution is above $85°C$, titrate with potassium permanganate. Use about 15 mL potassium permanganate solution. Add 10 mL sulfuric acid and 100 mL distilled water to a 1-L beaker to create a blank, and retitrate with potassium permanganate. Repeat titrations with the permanganate and blank. Calculate the normality of potassium permanganate as follows:

$$\text{Normality } (N) \text{ of } KMnO_4 = \frac{(\text{mg } Na_2C_2O_4)}{[(A \text{ mL}) - (B \text{ mL})] \times 0.000067}$$

Where:

A = mL of $KMnO_4$ used for sodium oxalate titration
B = mL of $KMnO_4$ used for distilled water

Average the results from several titrations to determine normality.

The equation below will determine the volume of $KMnO_4$ from Step 1 required to make a 50 mg/mL Mn standard solution in a 1 L flask.

$$\text{mL } KMnO_4 = 4.55/ N$$

2. Dissolve 10 g bisulfite in 100 mL distilled water.

3. Add volume of potassium permanganate stock solution calculated in step 2 to a 1-L volumetric flask and add 3 mL sulfuric acid. Add sodium bisulfite solution drop by drop to potassium permanganate solution until the pink color disappears. Bring entire volume to a boil. Allow volumetric flask to cool and bring volume back to 1 L with distilled water.

4. Prepare a series of manganese calibration standards. Add the volume of working manganese standard as shown in Table 17-1 to the 1,000-mL volumetric flask, then dilute to 1,000 mL with reagent-grade water.

Table 17-1   Preparing manganese calibration standards

Working Stock Manganese Solution	Calibration Standard
mL	μg/L as Mn
0	0 (Blank)
1	50
2	100
4	200
5	250
8	400
10	500
15	750
20	1,000

5. Add 100 mL sample to a volumetric flask. Add 100 mL of the series of standards selected to volumetric flasks. For high concentrations of manganese (greater than 1 mg/L), use a smaller volume of sample (50–75 mL). Add enough distilled water to increase the volume to 90 mL and skip step 8.

6. Add 1 drop hydrogen peroxide to sample and standards. Bring each flask to a boil and concentrate to 100 mL.

7. Add 1 g ammonium persulfate to sample and standards and bring each flask to a boil for no longer than 1 min. Allow samples to cool in a water bath for 1 min. Add distilled water to bring volumes to 100 mL.

8. Measure the color using the photometer or spectrophotometer with a filter or setting at 525 nm.

9. Create a plot of μg/L of manganese versus absorbance. The calibration curve should bracket the manganese concentration being measured. The curve should contain a blank and at least three other points.

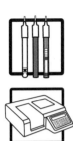

Chapter **18**

Nitrate

PURPOSE OF TEST

The determination of nitrate is desirable because nitrate in the body can be reduced to nitrite, which reacts with hemoglobin to form methemoglobin and can, in infants, lead to asphyxia. In an effort to control this problem, the US Environmental Protection Agency (USEPA) has imposed a maximum contaminant level of 10 mg/L in finished drinking waters (Pontius 1995).

LIST OF SIMPLIFIED METHODS

Only USEPA-approved methods may be used for Safe Drinking Water Act (SDWA) monitoring. Approved methods may be found in *Standard Methods for the Examination of Water and Wastewater*, Section 4500–NO_3^- D. Nitrate Electrode Method, E. Cadmium Reduction Method, and F. Automated Cadmium Reduction Method. The nitrate electrode and ultraviolet spectrophotometric screening methods are discussed in this chapter.

SIMPLIFIED PROCEDURES

Nitrate Electrode Method

The nitrate (NO_3^-) ion electrode is a selective sensor that responds through a porous, thin, inert membrane to ion activity in the concentration range of 0.14 mg/L to 1,400 mg/L and a pH of 3 to 9.

Warnings/Cautions.

Chloride and bisulfate ions in high concentrations can interfere with accuracy. Also, because the electrode responds to nitrate activity, ionic strength and pH must be kept constant to avoid erratic responses. A buffer minimizes this effect and complexes, or ties up, organic acids.

Apparatus.

- a pH meter with an expanded millivolt capability with resolution to 0.1 mV

- a double-junction reference electrode, the outer chamber filled with an ammonium sulfate [(NH$_4$)SO$_4$] solution

- a nitrate ion (NO$_3^-$) electrode (follow the manufacturer's directions for handling and storage)

- magnetic stirrer and TFE-coated stirring bars

Reagents.

Nitrate-free water. Use for all solutions, dilutions, and for rinsing the electrodes. Use redistilled or distilled, deionized water of the highest purity.

Potassium nitrate solution. Dry potassium nitrate (KNO$_3$) in an oven at 105° C for 24 hr. Dissolve 0.7218 g in nitrate-free water and dilute to 1 L. 1 mL = 100 mg NO$_3^-$ as nitrogen. Preserve with 2 mL chloroform per liter. This solution is stable for 6 months.

Standard nitrate solutions. Prepare by diluting 1.0, 10.0, and 50.0 mL of the stock standard solution to 100 mL with water to obtain standard solutions of 1.0, 10.0, and 50.0 mg/L NO$_3^-$ as nitrogen, respectively.

Buffer solution. Prepare by dissolving 17.32 g aluminum sulfate octadecahydrate [Al$_2$(SO$_4$)$_3 \cdot$ 18H$_2$O], 3.43 g silver sulfate (Ag$_2$SO$_4$), 1.28 g boric acid (H$_3$BO$_3$), and 2.52 g sulfamic acid (H$_2$NSO$_3$H) in approximately 800 mL nitrate-free water. Adjust pH to 3.0 by slowly adding a solution of 0.1N sodium hydroxide (NaOH), checking pH periodically. When adjusted to pH 3.0, dilute to 1 L in a volumetric flask. Transfer to a dark glass bottle for storage.

Sodium hydroxide solution, 0.1N. Carefully dissolve 4 g sodium hydroxide (NaOH) into 1 L nitrate-free water. Store this solution in a polyethylene bottle. Refer to *Standard Methods* for additional precautions.

Reference electrode solution. Used to fill the reference electrode, this solution may be prepared by dissolving 0.53 g ammonium sulfate [(NH$_4$)$_2$SO$_4$] in water and diluting to 100 mL.

Procedure.

1. Prepare a calibration curve by transferring 10 mL of the 1.0 mg/L nitrate standard to a 50-mL beaker. Add 10 mL buffer solution. Add a TFE-coated stir bar and place on the magnetic stirrer. Immerse the tips of the electrodes and ensure proper operation of the stir bar without hitting the electrodes. Turn the magnetic stirrer to about half speed and record the millivolt reading when stable, after approximately 1 min.

2. Remove the electrodes, rinse, and carefully blot dry. Repeat this for the 10 mg/L and 50 mg/L nitrate standards. Plot the millivolt potential against the nitrate concentration on semilogarithmic graph paper, with the nitrate concentration on the logarithmic axis and the millivolt potential on the linear axis. A straight line with a slope of 57 ± 3 millivolt/decade at 25° C should result. Recalibrate the electrodes several times a day by checking the 10 mg/L standard and adjusting the calibration control to the reading originally recorded on the calibration curve. For more information on

preparing a standard curve, see the Standard Calibration Curves section in chapter 1.

3. To determine the concentration of a sample, transfer 10 mL sample to a 50-mL beaker and add 10 mL buffer solution. Add a TFE-coated stir bar to the beaker and place on the magnetic stirrer. Immerse the electrodes and turn the stirrer to half speed. After approximately 1–2 min when the reading stabilizes, record the millivolt potential. Measure samples and standards at about the same temperature. Read the concentration from the calibration curve. A precision of approximately 2.5 percent is expected.

Ultraviolet Spectrophotometric Screening Method

This technique should be used only as a screen for samples with a low organic matter content.

Warnings/Cautions.

Analyze samples as soon as possible. If storage is necessary, store as long as 24 hr at 4°C. For longer storage, preserve the sample with 2 mL concentrated sulfuric acid per liter distilled water and store at 4°C.

NOTE: **When a sample is preserved with acid, nitrate and nitrite cannot be determined as individual species.**

Dissolved organic materials, surfactants, nitrite (NO_2^-), and chromium (Cr^{6+}) interfere with this method. Various inorganic ions not normally found in natural water, such as chlorite and chlorate, also may interfere. Inorganic substances can be compensated for by independent analysis of their concentrations and preparations of individual correction curves. Filtration of turbid samples may be necessary.

Apparatus.

*   a spectrophotometer for use at 220- and 275-nm wavelength with matched silica cells of 1-cm or longer light path

Reagents.

Nitrate-free water. Use redistilled or distilled, deionized water of highest purity to prepare all solutions and dilutions.

Stock nitrate solution. Dry potassium nitrate (KNO_3) in an oven at 105°C for 24 hr. Dissolve 0.7218 g in water and dilute to 1 L. 1 mL = 100 µg NO_3^- as nitrogen. Preserve with 2 mL chloroform/L. This solution is stable for 6 months.

Standard nitrate solution. Dilute 100 mL stock nitrate solution to 1 L with water. 1 mL = 10.0 mg NO_3^- as nitrogen. Preserve with 2 mL chloroform/L.

Hydrochloric acid solution, 1N. Add 83 mL of concentrated hydrochloric acid (HCl) to 1 L distilled, nitrate-free water.

Procedure.

1. To 50 mL clear sample (filtered, if necessary), add 1 mL hydrochloric acid solution and mix thoroughly.

2. Prepare a standard curve in the range of 0 to 7 mg NO_3^- as nitrogen/L by diluting to 50 mL the following volumes of the standard nitrate solution:

0, 1.00, 2.00, 4.00, 7.00, ... 35.0 mL. Treat the nitrate standards in the same manner as the samples.

3. Read absorbance or transmittance against redistilled water set at zero absorbance or 100 percent transmittance. Use a 220-nm wavelength to obtain the nitrate reading and a 275-nm wavelength to determine the interference caused by dissolved organic matter.

4. For samples and standards, subtract two times the absorbance reading at 275 nm from the reading at 220 nm to obtain absorbance caused by nitrate. Construct a standard curve with standards by plotting absorbance against concentration. Using corrected sample absorbancies, obtain sample concentrations directly from the standard curve. For more information on preparing a standard curve, see the Standard Calibration Curves section in chapter 1.

NOTE: If the correction value is more than 10 percent of the reading at 220 nm, do not use this method.

Chapter **19**

Nitrite

PURPOSE OF TEST

The determination of nitrite is desirable because nitrate in the body can be reduced to nitrite. Nitrite reacts with hemoglobin to form methemoglobin and can, in infants, lead to asphyxia. In an effort to control this problem, the US Environmental Protection Agency (USEPA) has imposed a maximum contaminant level of 10 mg/L in finished drinking waters (Pontius 1995).

LIST OF SIMPLIFIED METHODS

Colorimetric Method

Only USEPA-approved methods may be used for Safe Drinking Water Act (SDWA) monitoring. Approved methods may be found in the current edition of *Standard Methods for the Examination of Water and Wastewater*, Sections 4500–NO_2^- and 4110. Ion Chromatography. See also USEPA Method 300.0, Determination of Inorganic Anions in Water by Ion Chromatography.

SIMPLIFIED PROCEDURE

Nitrite is determined by forming a reddish purple azo dye produced at a pH between 2.0 and 2.5 by coupling diazotized sulfanilamide with N-(1-naphthyl)-ethylene-diamine (NED) dihydrochloride. The applicable range for this spectrophotometric method is 10 to 1,000 mg/L NO_2^- as nitrogen. Higher NO_2^- concentrations can be determined by diluting the sample.

Colorimetric Method

Warnings/Cautions.

This method is susceptible to interference from free chlorine (Cl_2) and nitrogen trichloride (NCl_3). These interferences would probably not coexist with NO_2^-. Nitrogen trichloride (NCl_3) will give a false red color to the sample in this test. Additional ions that may interfere may be found in *Standard Methods*, Sec. 4500–NO_2^- B.1. Colored ions that alter the color system may also interfere. Suspended solids should be removed by filtration.

Never acidify samples. Perform this analysis as soon as possible to prevent bacterial conversion of NO_2^- to nitrate (NO_3^-) or to ammonia (NH_3). Slow this conversion by keeping the sample at $4°C$. To store samples from 1 to 2 days 48 hr holding time at $40°C$, freeze at $-20°C$.

Apparatus.

- a spectrophotometer with a light path of 1 cm or longer that can operate at 543 nm or a filter photometer, with a light path of 1 cm or longer, equipped with a green filter and with maximum transmittance near 540 nm.

Reagents.

Nitrite (NO_2^-)-free water. Perform all solutions, dilutions, and rinses with nitrite-free water. If the quality of the reagent water is not known or is suspect, nitrite-free water may be prepared by one of the following procedures:

1. Add one small crystal of potassium permanganate ($KMnO_4$) and either barium hydroxide [$Ba(OH)_2$] or calcium hydroxide [$Ca(OH)_2$]. Redistill in an all-glass apparatus, discarding the first 50 mL and collecting the fraction that is free of the potassium permanganate. A red color with DPD reagent indicates the presence of potassium permanganate. Test small volumes frequently.

2. Add 1 mL concentrated sulfuric acid (H_2SO_4) and 0.2 mL manganese sulfate ($MnSO_4$) solution (36.4 g $MnSO_4 \cdot H_2O$/100 mL distilled water) to each liter distilled water, and add 1 to 3 mL potassium permanganate solution (400 mg $KMnO_4$/L distilled water), enough to turn the solution pink. Redistill as described previously.

Color reagent. Prepare by adding 100 mL of an 85 percent phosphoric acid (H_3PO_4) solution and 10 g sulfanilamide ($C_6H_8N_2O_2S$) to 800 mL nitrite-free water. After dissolving the sulfanilamide completely, add 1 g NED dihydrochloride. Mix to dissolve and dilute to 1 L with water. This solution is stable for 1 month when refrigerated in a dark bottle.

Stock nitrite solution. Prepare from reagent grade sodium nitrite ($NaNO_2$). Dissolve 1.232 g $NaNO_2$ in nitrite-free water and dilute to 1 L. 1 mL = 250 mg of NO_2^- as nitrogen. Preserve with 1 mL chloroform.

NOTE: **Nitrite is readily oxidized in the presence of moisture. Always keep the reagent bottle tightly closed when not in use, and note the value generated from the undiluted material when first opened. When there is appreciable deviation, use a fresh bottle of sodium nitrite.**

Intermediate nitrite solution. Dilute 50 mL stock nitrite solution to 250 mL with water. 1 mL = 50 mg of NO_2^- as nitrogen. This solution needs to be prepared each day the analysis is performed.

Standard nitrite solution. Dilute 10.00 mL intermediate nitrite solution to 1 L with water. 1 mL = 0.5 mg of NO_2^- as nitrogen. This solution needs to be prepared each day the analysis is performed. For more information on preparing a standard curve, see the Standard Calibration Curves section in chapter 1.

Procedure.

1. Remove any suspended solids present by filtration through a 0.45-μm diameter pore membrane filter.

2. Determine the pH of the sample. Adjust the pH of the sample to the range of 5 to 9 with $1N$ hydrochloric acid (HCl) or $1N$ ammonium hydroxide (NH$_4$OH), as needed.

3. To develop the color of the sample or standard, add 2 mL color reagent to a 50-mL portion of sample. Dilute the sample, if necessary, and mix.

4. Between 10 min and 2 hr after adding the color reagent to both the samples and standards, using the spectrophotometer, measure the absorbance of the solutions at 543 nm. Record each measurement. Plot the absorbance of nitrite standards, as nitrogen, against the concentration of each.

The solutions and concentrations provided in Table 19-1 can be used for a standard curve.

Compute the sample concentration directly from the curve.

Use the light paths in Table 19-2 for the indicated nitrite concentrations as a guide.

Table 19-1   Solutions and concentrations for nitrite standards

mL of Standard Solution 1.0 mL = 0.5 mg of NO$_2^-$N	Concentration When Diluted to 50 mL X mg/L of NO$_2^-$N
0.0	Blank
1.0	0.01
2.0	0.02
4.0	0.04
6.0	0.06
8.0	0.08
10.0	0.10
20.0	0.20

Table 19-2   Light paths to determine nitrite concentrations

Light Path Length cm	NO$_2^-$ as Nitrogen $\mu g/L$
1	2–25
5	2–6
10	<2

This page intentionally blank

Chapter **20**

Ozone (Residual)

PURPOSE OF TEST

Ozone is an effective germicide used as a disinfectant in water treatment. It is also used as an oxidizing agent to destroy organic compounds that produce taste and odor in drinking water. Ozone reacts with organic coloring matter and oxidizes reduced iron and manganese to insoluble oxides. The tests described in this chapter can be used to measure residual ozone. Continuous monitors are generally used to measure the ozone added to the water.

LIST OF SIMPLIFIED METHODS

A variety of colorimetric tests have been used to measure residual ozone, but most tests are nonspecific and actually measure total oxidant levels. The preferred procedure is the indigo method, which responds rapidly to ozone and is subject to few interferences. In addition, test kits based on the indigo method are available. For more information, see *Standard Methods for the Examination of Water and Wastewater*, Section 4500–O_3.

Test kits based on the DPD method are also available, but they will not be described in this chapter. The DPD method gives similar results (Gordon 1987), but the reading must be performed within 5 min. Samples analyzed following the indigo procedure are stable for as long as 4 hours.

SIMPLIFIED PROCEDURE

The indigo colorimetric procedure is quantitative and selective. It is applicable to lake water; river infiltrate; manganese-containing and extremely hard groundwater; and biologically treated domestic wastewaters. In acidic solution, ozone rapidly removes the color from indigo. The decrease in absorbance is linear as concentration increases.

Indigo Colorimetric Procedure

Warnings/Cautions.

Pay careful attention to sample collection. Ozone is a gas and is easily lost from water. Collect the sample to minimize aeration, and fill the ampoule immediately.

Manganese can interfere with the indigo procedure. If manganese is present, use the water to be tested as a blank after the ozone has been removed by shaking it vigorously for at least 10 sec or by adding glycine reagent to destroy the ozone as detailed in *Standard Methods*, Section 4500–O$_3$. Chlorine does not interfere with this procedure.

Caution: Reagents used for this test degrade over time. Order only a 3-month supply. Date the test kits when received and discard unused reagents after 6 months. Store vacuum-sealed vials in a cool, dry place away from direct sunlight.

Caution: Ozone has an Occupational Safety and Health Administration (OSHA) Permissible Exposure Limit of 0.1 ppm (0.2 mg/m^3). It is a potential inhalation hazard and can be irritating to the eyes, nose, and mucous membranes.

Apparatus.

- a colorimeter for use at 600 ± 5 nm

- beakers for collecting a 40-mL sample

Reagents.

The reagents required for this analysis are supplied in air-evacuated ampoules. Each sample requires two ampoules, one for the sample and one for the blank. The same blank may be used for all samples analyzed within 4 hr of its preparation.

Procedure.

1. Collect at least 40 mL ozone-free water in a beaker or other open container for the blank.

2. Collect at least 40 mL sample in a beaker. Take the sample so there is no aeration and no bubbles are formed during collection. Do not shake the sample, and fill the ampoule immediately.

3. Fill one ampoule with sample and one with the blank by immersing the tip well below the surface of the liquid and breaking off the tip by pressing against the side or the bottom of the container. Allow the ampoule to fill, keeping the break well below the surface while it is filling to prevent air from being drawn into the ampoule.

4. Invert the ampoule several times to dissolve the reagent powder, and wipe any liquid or fingerprints off its outside. The blue color will be bleached if ozone is present.

5. Adjust the colorimeter according to the manufacturer's recommendations. Place the blank ampoule into the colorimeter and set the reading for 0 mg/L ozone. Remove the blank ampoule and insert the sample ampoule. Read the colorimeter and follow the manufacturer's instructions.

This page intentionally blank

Chapter **21**

pH

PURPOSE OF TEST

The purpose of this test is to determine the relative acidity or alkalinity of a sample. The pH scale runs from 0 to 14. Water with a pH of 7.0 is considered neutral. Acidic water has a pH below 7.0, and an alkaline water has a pH above 7.0. A water's pH is determined by its hydrogen ion activity. This measurement is important in drinking water treatment because it affects process controls such as chlorination, coagulation, softening, and corrosion. The pH test helps the plant operator maintain proper chemical doses.

LIST OF SIMPLIFIED METHODS

Refer to *Standard Methods for the Examination of Water and Wastewater* Section 4500–H$^+$ pH Value. See also the *USEPA Methods for Water and Wastewater*, Number 150.1. For more extensive information, see Bates (1978).

Inexpensive portable test devices are available from many manufacturers (see appendix A).

SIMPLIFIED PROCEDURE

pH Value

Warnings/Cautions.

The glass electrode used in this technique is usually not subject to solution interferences from color, turbidity, colloidal matter, oxidants, reductants, or high salinity.

Coatings of oily material or particulate matter can impair electrode response. These coatings can be removed by gentle wiping or detergent washing followed by distilled water rinsing. An additional treatment with 10 percent hydrochloric acid

(HCl) may be necessary to remove any film. Follow specific electrode cleaning instructions provided by the manufacturer.

Temperature effects on electrode measurements of pH depend on the reference electrode used, pH of the solution within the electrode, and the pH of the test solution (Hach 1997). Effects can be controlled by keeping the sample and calibration solutions at the same temperature, usually room temperature. Always note the temperature of both the calibration solution and the sample. Alternatively, use an automatic temperature compensation probe. The temperature still must be recorded, even when this probe is used.

Do not stir groundwater samples while they are open to the atmosphere; degasification will cause significant changes in pH. Insert the probe through a stopper and fill the sample container to exclude air space before stirring and reading pH.

Apparatus.

- a pH meter, laboratory or field model, accurate and reproducible to 0.1 pH units and equipped with temperature compensation adjustment

- glass or epoxy plastic electrode. Measuring and reference electrodes are required. However, combination electrodes that include both measuring and reference functions are available with a solid gel-type filling material and require little maintenance.

- beakers, polyethylene or TFE

- a TFE magnetic stirring bar

- thermometer or automatic temperature sensor for compensation

- electrode(s) must be stored in a storage solution. Follow the manufacturer's recommendations.

Reagents.

Secondary standard buffer solutions. Traceable to National Institute of Standards and Technology, salts are available commercially. These validated standards are recommended for routine use.

Procedure.

1. Remove the electrode from the storage solution, rinse with distilled water, and blot dry with a soft tissue. Detailed calibration and operating instructions are included with each manufacturer's meter. The analyst should be familiar with individual meter requirements.

2. In a beaker of sufficient volume to allow the addition of a small TFE-coated stir bar and immersion of the electrode, add the initial buffer. Buffers should bracket the pH of the samples being tested. Discard buffers after each use. Place on a magnetic stirrer and stir gently. Adjust the meter to read the pH of the buffer as the initial point. Remove the electrode and rinse thoroughly with distilled water. Immerse in the second buffer solution. Set the meter dial to read the pH of the second buffer. Repeat a third time, with a buffer below 10 and approximately 3 pH units different from the second. The pH measurement should be within 0.1 units of the stated value. If not, there may be a problem with the electrode or meter.

3. Immerse the electrode in the solution while slowly stirring the sample. Read the pH. Remove the electrode and repeat with a second fresh volume of sample. The pH of the two solutions should read within 0.1 pH units. If not, the electrode may need to be conditioned by briefly immersing it in the sample and reading the pH of a second fresh solution.

4. Analyze samples as soon as possible. There is no holding time for this method.

This page intentionally blank

Chapter **22**

Phosphate

PURPOSE OF TEST

Phosphate is seldom present in significant concentrations in natural waters used for drinking purposes but may be added during the treatment process as orthophosphate (simple phosphate salts) or condensed phosphate (meta-, pyro-, and polyphosphate). Phosphates are sometimes applied to water in small doses to prevent or delay iron precipitation or corrosion and to control the deposition of calcium carbonate scale throughout a distribution system. Phosphates are also used extensively for treating boiler waters.

This chapter is divided into two procedures. The first procedure, orthophosphate, is used when orthophosphates are added in the treatment process: the second, total phosphate, is suitable when condensed phosphates are used. The total phosphate analysis is performed in two steps: (1) a digestion procedure that converts all phosphates to orthophosphate; and (2) a colorimetric determination of the orthophosphate concentration.

Both procedures involve the colorimetric determination of orthophosphate with the use of either a spectrophotometer or, less frequently, a photometer. Colorimetric laboratory equipment measures variations in color too small to be seen by the human eye. The color variations are caused by differences in orthophosphate concentrations and measured by reading the percent transmittance or absorbance of light passed through the processed sample.

LIST OF SIMPLIFIED METHODS

Refer to *Standard Methods for the Examination of Water and Wastewater*, Sec. 4500–P., A., and B. for information on selecting a method and preparing a sample.

SIMPLIFIED PROCEDURES

Orthophosphate Determination

Warnings/Cautions.

These procedures are suitable for clear and colorless waters. For brines with high chloride content and waters that contain more than 1 mg/L of ferric iron, nitrite, or oxidizing agents, such as chromate, consult *Standard Methods*.

Never use commercial detergents that contain phosphate to clean glassware used in phosphate analysis.

Errors can also result from condensed moisture, dust, or fingerprints on the cuvette surfaces, or bubbles within the solution.

Be alert to the emergence of off-colors and turbidity in the samples and standards that can easily pass unnoticed during photometric measurements. Verify every unusual value to remove suspicion about its validity. Be alert to the possibility of fatigue and loss of sensitivity in the photocells, fluctuations in the line voltage, and electrical and mechanical failures in the instrument. If loss of sensitivity is observed, consult the manufacturer's instructions.

Prepare a reagent blank along with every set of samples and at least one standard in the upper end of the optimum concentration range to confirm the reliability of the calibration curve, reagents, instrument, and technique of the analyst. When instruments are used for hourly readings, they should be checked once or twice during every shift.

Photometers yield the best results in their mid-operating range, represented by a 20 percent transmittance or 0.70 absorbance on the lower end of the scale and an 80 percent transmittance or 0.10 absorbance at the upper end. At low transmittance (10 percent) or high absorbance (1.0), the instrument responds insensitively to appreciable color and concentration changes. At high transmittance (95 to 100 percent) or low absorbance (0.02), slight construction differences in the cuvettes (sample containers or cells) and their placement in the sample compartment can affect the readings.

Apparatus.

To determine orthophosphate and total phosphate, one of the following types of photometric equipment is required:

- a spectrophotometer, for use at a wavelength of 650, 660, or 880 nm that provides a light path of 1 cm or longer; or a filter photometer, equipped with a filter that has maximum transmission in the wavelength range of 400 to 470 nm that provides a light path of 1 cm or longer

In addition to photometric equipment, the following supplies are required to determine orthophosphate. Acid wash all glassware to ensure no phosphate residues are present.

- medicine droppers to add stannous chloride reagent ($SnCl_2$), phenolphthalein, and strong acid solution

- a 100-mL graduated cylinder and volumetric pipets to measure standards, samples, and other reagents

- seven or more 250-mL volumetric flasks
- seven or more 250-mL Erlenmeyer flasks

Reagents.

All reagents listed are required for orthophosphate and total phosphate determinations unless otherwise specified.

Stock phosphate solution.

1. On an analytical balance, weigh 0.2195 g anhydrous potassium dihydrogen phosphate, also called potassium monobasic phosphate (KH_2PO_4).

2. Carefully transfer to a 250-mL beaker and dissolve in 100 mL distilled water.

3. Transfer the solution to a 1-L volumetric flask. Rinse beaker with three 100-mL portions of distilled water. Add rinsings to flask.

4. Dilute to the 1 L mark with distilled water. Stopper and mix thoroughly. 1 mL = 0.05 mg PO_4^{3-} as P. This solution is available commercially.

Standard phosphate solution. With a volumetric pipet, measure carefully 10 mL stock phosphate solution into a 1-L volumetric flask. Dilute to the 1 L mark with distilled water. Stopper and mix thoroughly. 1 mL = 0.0005 mg PO_4^{3-} as P.

Phenolphthalein indicator solution. Weigh 0.5 g phenolphthalein disodium salt powder and dissolve in 100 mL distilled water.

Strong acid solution.

1. With a 1-L graduated cylinder, measure 600 mL distilled water and pour into a 1,500-mL beaker.

2. With a 500-mL graduated cylinder, measure 300 mL concentrated sulfuric acid (H_2SO_4).

3. While stirring constantly, slowly add sulfuric acid to the distilled water. Considerable heat is generated by mixing sulfuric acid and distilled water, so pour slowly and mix well to avoid dangerous spattering.

4. Allow the solution to cool to room temperature.

5. Add 4.0 mL concentrated nitric acid (HNO_3) to the cooled solution and mix well.

6. Transfer to a 1-L graduated cylinder and dilute to the 1 L mark with distilled water. Mix thoroughly by pouring back into the beaker and stirring.

Acid-molybdate solution.

1. With a 500-mL graduated cylinder, measure 400 mL distilled water and pour into a 1,500-mL beaker.

2. With a 500-mL graduated cylinder, measure 280 mL concentrated sulfuric acid (H_2SO_4).

3. While stirring constantly, slowly add sulfuric acid to the distilled water. Considerable heat is generated by mixing sulfuric acid and distilled water, so pour slowly and mix well to avoid dangerous spattering.

4. Allow the solution to cool to room temperature.

5. Weigh 25 g ammonium molybdate [$(NH_4)_6Mo_7O_{24} \cdot 4H_2O$]. Carefully transfer to a 250-mL beaker and dissolve in 175 mL distilled water.

6. Add the molybdate solution to the cooled acid solution, dilute to 1 L with distilled water, and mix thoroughly.

Stannous chloride reagent.

1. Weigh 2.5 g stannous chloride dihydrate ($SnCl_2 \cdot 2H_2O$) and place in a 250-mL heat-resistant glass bottle.

2. With a graduated cylinder, measure 100 mL reagent-grade glycerol or glycerine and add to the bottle. Place the bottle in a hot water bath and stir the chemicals with a glass rod until the stannous chloride completely dissolves.

In addition, the following reagents are required to determine total phosphate.

Sulfuric acid solution.

1. With a 1-L graduated cylinder, measure 600 mL distilled water and pour into a 1,500-mL beaker.

2. With a 500-mL graduated cylinder, measure 300 mL concentrated sulfuric acid (H_2SO_4).

3. While stirring constantly, slowly add sulfuric acid to distilled water. Considerable heat is generated by mixing sulfuric acid and distilled water, so pour slowly and mix well to avoid dangerous spattering.

4. Allow the solution to cool to room temperature.

5. Transfer to a 1-L graduated cylinder and dilute to the 1 L mark with distilled water. Mix thoroughly.

Ammonium persulfate [$(NH_4)_2S_2O_8$] or *potassium persulfate* ($K_2S_2O_8$).

Sodium Hydroxide (1*N*). Weigh 40 g sodium hydroxide (NaOH) pellets. Dissolve slowly in 1 L distilled water, stirring constantly. Place beaker under fume hood, surrounded by running water or ice water, while preparing solution.

Procedure.

Operating precautions for photometric methods. Several precautions are advisable for operating all photometric instruments. Zero and standardize each instrument according to the manufacturers' directions. Keep the cover on the sample compartment tightly closed while measuring to minimize light leakage. Keep the sample compartment clean and dry at all times to prevent impairment and corrosion of the vital units. Wipe the external surfaces of the cuvette clean with a soft cloth or cleansing tissue to remove dust, liquid, and fingerprints that can cause incorrect readings. Rinse cuvettes free of remains from previous tests to ensure reliable readings of the standards and samples under examination. It may be necessary to rinse with distilled water to achieve the desired removal. Instruments' readout systems vary, so maintain the proper eye position to make an accurate reading for a nondigital readout.

Cuvette evaluation.   The placement of the cuvettes in the sample compartment requires special attention. Conduct the evaluation in the following manner.

1. After turning the instrument on and allowing it to warm up for several minutes, set the wavelength control or insert the filter in the proper slot.

2. Fill each cleaned cuvette to the same level with distilled water. Wipe external surfaces with a soft, lint-free cloth to remove any dust, liquid, or fingerprints.

3. Place the first cuvette into the sample compartment and adjust the meter to a 95.0 percent transmittance or 0.020 absorbance reading.

4. Remove the cuvette and replace it with another cuvette. Record the transmittance or absorbance reading of each cuvette.

5. If necessary, rotate the cuvette in the sample compartment so all readings are the same (95.0 percent transmittance or 0.020 absorbance). A difference of 0.5 percent transmittance or 0.002 absorbance in readings between cuvettes is acceptable.

6. Permanently mark each cuvette to indicate the proper placement in the cuvette holder necessary to achieve the 0.5 percent transmittance or 0.002 absorbance tolerance.

Prepare the standard curve and determine orthophosphate (stannous chloride colorimetric method).

For more information on preparing a standard curve, see the Standard Calibration Curves section in chapter 1.

1. Using Table 22-1, prepare a series of phosphate standards by measuring the indicated volumes of standard phosphate solution into separate 100-mL volumetric flasks. Add distilled water to the 100-mL mark and mix. Transfer standards to 250-mL Erlenmeyer flasks.

2. Place 100-mL sample (should be clear and colorless) in a 250-mL flask.

Table 22-1   Dilutions to prepare phosphate standards

Standard Phosphate[*] Solution Added to Flask *mL*	Phosphate[*] Concentration in 100-mL Flask *mg/L*
0	0
1	0.5
2	0.10
3	0.15
4	0.20
5	0.25
6	0.30

*Phosphate expressed as phosphorus.

NOTE: Rate of color development and intensity of color are temperature dependent. Hold samples and standards within 2° C of each other in the range of 20° to 30° C.

3. Add 1 drop phenolphthalein indicator to sample flask and mix. If the sample turns pink, add a strong acid solution drop by drop to discharge the pink. If more than 5 drops are required, take a smaller sample and dilute to 100 mL with distilled water.

4. Add 4 mL acid-molybdate solution to each flask and mix thoroughly.

5. Add 10 drops stannous chloride solution to each flask and mix thoroughly.

6. Allow color to develop for 10 min. After 10 min, but before 12 min (using the same specific intervals for all determinations), measure color photometrically.

7. Pour an adequate volume of each developed standard and sample into separate, clean, matched cuvettes.

8. Set spectrophotometer wavelength to 690 nm or place a 400 to 470 nm filter into the filter photometer.

9. After sufficient warmup time, zero the instrument with the reagent blank (0 mg/L phosphate standard).

10. Record the absorbance or percent of transmittance meter reading for each standard and sample. If the meter is not digital, adopt the correct eye position while making the reading.

NOTE: Ensure the readings are made with the sample compartment tightly closed.

11. Plot the instrumental readings of the standards versus concentration in one of two ways to obtain the calibration curve: (1) Plot the absorbance versus concentration on ordinary coordinate graph paper; (2) Plot the percent transmittance on the logarithmic scale versus the concentration on the linear scale of semilogarithmic graph paper. Draw a smooth curve to connect the points. A straight line (through the origin of the graph) indicates an ideal color system (confirming Beer's law) for photometric use. See chapter 1 for a typical calibration curve and conversion of percent transmittance to absorbance.

Although calibration curves tend to deviate from straight lines at high concentrations, other causes of deviation may originate in instrumental factors such as broad band width of the color filter, stray light caused by light leaks, optical imperfections, or improper optical alignment or maintenance. See Table 1-3 for the conversion from percent transmittance to absorbance readings if semilogarithmic graph paper is unavailable.

12. To determine the amount of orthophosphate in the sample, compare the photometric reading (absorbance or percent transmittance) for the sample with the standard calibration curve and extrapolate the concentration.

Total Phosphate Determination

Prepare the standard curve and determination of total phosphate (Persulfate Digestion and Stannous Chloride Colorimetric Method). For more information on preparing a standard curve, see the Standard Calibration Curves section in chapter 1.

Refer to *Standard Methods for the Examination of Water and Wastewater*, Section 4500–P., B (5.) Digestion for T.P.

Apparatus.

One of the following types of photometric equipment is required to determine total phosphate.

- spectrophotometer, for use at a wavelength of 650, 660, or 880 nm providing a light path of 1 cm or longer or a filter photometer equipped with a filter that has maximum transmission in the wavelength range of 400 to 470 nm and provides a light path of 1 cm or longer

The following supplies are also required to determine total phosphate. Acid wash all glassware to ensure no phosphate residues are present.

- medicine droppers to add stannous chloride reagent ($SnCl_2$), phenolphthalein, and strong acid solution

- a 100-mL graduated cylinder and volumetric pipets to measure standards, samples, and other reagents

- seven or more 250-mL volumetric flasks

- medicine dropper for dispensing sodium hydroxide solution

- gas burner or electric hot plate

- one or more wire gauze squares, 20-mesh

- four or more glass beads

- one or more glass stirring rods

- wash bottle for rinsing flask, beads, and stirring rod

Reagents.

See reagents in Orthophosphate Determination section.

Procedure.

1. Using Table 22-2, prepare a series of phosphate standards according to previous page.

Table 22-2   Sample volumes for indicated metaphosphate range

Sample Volume mL	Metaphosphate* range mg/L
100	0.1–2.0
50	2.1–4.0
25	4.1–6.0

*Phosphate expressed as phosphorus.

2. Measure the sample volume for the indicated metaphosphate range.

If a sample of only 50 mL is needed, add 50 mL distilled water to bring the total volume to 100 mL. If a sample of only 25 mL is needed, add 75 mL distilled water. Measure the additional distilled water with a graduated cylinder. Place the sample (and extra distilled water, if needed) in a 250-mL flask.

3. Add 1 drop phenolphthalein indicator solution to each flask.

4. If a red or pink color develops, add sulfuric acid solution drop by drop to discharge the color. Then add an additional 1 mL sulfuric acid solution.

5. Add 0.4 g ammonium persulfate or 0.5 g potassium persulfate to each flask.

6. Boil gently for 30 to 40 min or until a final volume of 10 mL is reached. Prevent spattering and bumping of the boiling liquid by placing a wire gauze over the electric or gas heat source and resting the flask on the wire gauze. For further protection, add four or five glass beads to the flask or place a glass stirring rod in the flask.

7. Cool the standards and samples to room temperature.

8. Dilute the standards and samples to 30 mL with distilled water.

9. Add 1 drop phenolphthalein indicator to each flask.

10. Add sodium hydroxide solution until a faint pink color appears.

11. Add enough distilled water to bring the volume to the 100-mL mark.

To complete the total phosphate determination, refer to steps 3–12 in the Orthophosphate Determination section in this chapter.

Refer to *Standard Methods for the Examination of Water and Wastewater*, Section 4500–P., A. and B. for information on selecting a method and preparing a sample.

Chapter **23**

Silica

PURPOSE OF TEST

Silicon is an extremely abundant element. It appears as the oxide (silica) in quartz and sand and combines with metals to form many complex silicate minerals. The silica content of natural waters ranges from 1 to 30 mg/L, although concentrations of 100 to 1,000 mg/L can be found. Silica in water is undesirable for many industrial uses because it forms silicate scales that are difficult to remove from equipment. Silica is most often removed by strongly basic anion-exchange resins in a deionization process, by distillation, or by reverse osmosis. Precipitation by magnesium oxide in either a hot or cold lime-softening process can also be used.

LIST OF SIMPLIFIED METHODS

Other methods are provided in *Standard Methods for the Examination of Water and Wastewater*, Section 4500–SiO_2, but they exceed the scope of this manual.

SIMPLIFIED PROCEDURE

Molybdosilicate Method

A well-mixed sample is filtered through a 0.45–μm membrane filter. When ammonium molybdate is added in an acidic solution, the filtrate will form a yellow complex proportional to the dissolved silica in the sample. The color complex can be measured with a spectrophotometer or by visual comparison. Approximately 1 mg/L may be detected with this method.

Warnings/Cautions.

Because apparatus and reagents may contribute silica, avoid using glassware as much as possible and employ reagents low in silica. Use a blank to correct for silica introduced from these sources.

Tannin, large amounts of iron, color, turbidity, sulfide, and phosphate interfere. Treatment with oxalic acid eliminates interference from phosphate and decreases interference from tannin. If necessary, use photometric compensation to correct for interference from color and turbidity.

Apparatus.

- a spectrophotometer for use at 410 nm, with a 1-cm or longer cell or a filter photometer with a violet filter having a maximum transmittance as near 410 nm as possible and a 1-cm or longer cell.

- 100-mL platinum dishes

- 50-mL Nessler tubes, matched, tall form.

Reagents.

Set aside and use batches of chemicals low in silica. Store all reagents in plastic containers.

Sulfuric acid, H_2SO_4, $1N$.

Hydrochloric acid, HCl, 1:1.

Ammonium molybdate reagent. Dissolve 10 g ammonium molybdate [$(NH_4)_6Mo_7O_{24} \cdot 4H_2O$] in distilled water in a 100-mL volumetric flask by gently stirring and warming. Dilute to volume. Filter if necessary. Adjust to a pH of 7 to 8 with silica-free ammonium hydroxide (NH_4OH) or sodium hydroxide (NaOH). Store in a plastic bottle.

Oxalic acid solution. Dissolve 7.5 g oxalic acid ($H_2C_2O_4 \cdot H_2O$) in distilled water in a 100-mL volumetric flask and dilute to volume. Store in a plastic bottle.

Stock silica solution. Dissolve 4.73 g sodium metasilicate nonahydrate ($NaSiO_3 \cdot 9H_2O$) in distilled water and dilute to 1 L.

Store in a tightly sealed plastic bottle. Alternatively, purchase a prepared silica stock standard from a commercial source.

Standard silica solution. Dilute 10.0 mL stock solution to 1 L with recently boiled and cooled distilled water. 1 mL = 10.0 μg SiO_2. Store in a tightly sealed plastic bottle.

All reagents are commercially available.

Procedure.

1. For color development, to 50.0 mL of sample rapidly add in succession, 1.0 mL 1+1 HCl, 2.0 mL ammonium molybdate reagent. Mix by inverting at least six times. Let stand for 5 to 10 min.

   Add 2.0 mL oxalic acid solution and mix thoroughly. Read the color after 2 min but before 15 min, measuring time from the addition of the oxalic acid solution. Measure the color either photometrically or visually.

2. If unreactive silica must be detected, see *Standard Methods*, Section 4500—Si D.

3. For color or turbidity correction, prepare a separate blank for each sample that requires such correction. Carry two identical portions of each sample through the entire procedure. To one portion add all the reagents as directed in step 1.

To the other portion add HCl and oxalic acid but no molybdate. Adjust the spectrophotometer or filter photometer to zero absorbance with the blank that contains no molybdate before reading the absorbance of the molybdate-treated sample.

4. Prepare a calibration curve using approximately six standards and a reagent blank. From Table 23-1, select the light path length and wavelength for the desired working range. For more information on preparing a standard curve, see the Standard Calibration Curves section in chapter 1.

    Set the spectrophotometer or filter photometer at zero absorbance with distilled water and read all standards, including a reagent blank, against distilled water. Plot the µg silica in the remaining 55 mL developed solution against photometer readings. Run a reagent blank and at least one standard with each group of samples to confirm that the calibration curve previously established has not changed.

5. If a visual comparison is desired, refer to either *Standard Methods* or the US Environmental Protection Agency (USEPA) manual *Methods for Analysis of Water and Wastes.*

6. Calculation

$$\text{mg SiO}_2/ \text{ L} = \frac{\mu g \text{ SiO}_2 (\text{in 55 mL final volume})}{\text{mL sample}}$$

Table 23-1   Selecting light path length for various silica concentrations

Light Path	Silica in 55 mL final volume, µg		
cm	410 nm	650 nm	815 nm
1	200–1,300	40–300	20–100
2	100–700	20–150	10–50
5	40–250	7–50	4–20
10	20–130	4–30	2–10

This page intentionally blank

Chapter **24**

Sodium

PURPOSE OF TEST

Sodium is present in most natural waters in concentrations of less than 1 mg/L to more than 500 mg/L. Concentrations of sodium greater than 500 mg/L where chloride is also present can give water a salty taste.

LIST OF SIMPLIFIED METHODS

The method described here uses a sodium electrode to determine concentration. The flame emission photometer, the flame atomic absorption, and the inductively coupled plasma methods are described in *Standard Methods for the Examination of Water and Wastewater*, Section 3500–Na, for laboratories where this equipment is available.

SIMPLIFIED PROCEDURE

Sodium Electrode Method

Warnings/Cautions.

The method described covers the concentration range of 10 mg/L to 1,000 mg/L. The sodium electrode can be used for concentrations outside this range, but additional calibration solutions will be needed. The stated range should meet process control needs.

Apparatus.

- laboratory pH/mV meter with readability to 0.1 mV

- sodium electrode

- reference electrode

- magnetic stirrer and stir bars

- graph paper, four-cycle semilogarithmic paper for preparing calibration curves

- beakers, 50-mL for 25-mL samples or 250-mL for 100-mL samples

- 1-mL and 10-mL volumetric pipets

- 50-mL and 500-mL graduated cylinders

- 100-mL volumetric flasks

Reagents.

Sodium and potassium ionic strength adjustor (ISA).
Sodium reference electrolyte solution. Available from electrode suppliers.
Sodium standards (10, 100, and 1,000 mg/L as sodium). Either purchase these standards individually or purchase the 1,000 mg/L solution and prepare the other concentration levels by adding 1.0 mL and 10.0 mL, respectively, to 100-mL volumetric flasks and fill to the mark with distilled or deionized water. Stopper the flasks and invert at least six times to thoroughly mix the solution.
Distilled or deionized water.
Electrode rinse solution. Add 10 mL of ISA solution to a 1-L squeeze bottle and fill it with distilled water.

Procedure.

1. Install the electrodes in the pH/mV meter following the manufacturer's instructions. See the electrode instruction manual to prepare, condition, and care for the electrode.

2. Using a volumetric pipet to measure the volume, transfer 25 mL of each calibration solution into a separate beaker. Add the sodium and potassium ISA to each beaker. Add either 2.5 mL ISA solution, or one powder pillow, or 0.8 g powder.

3. Add a stirring bar and use the magnetic stirrer to mix thoroughly. Stir the solution moderately while measuring. Place the electrodes into the calibration solution. Record the meter reading after it has stabilized. Analyze the other two calibration standards, and rinse the electrode with electrode rinse solution between measurements.

4. Plot the readings obtained on the semilogarithmic paper with the mV reading plotted on the normal axis versus the concentration of the calibration solution along the logarithmic axis. The difference in electrode response caused by the tenfold concentration difference should be 54 to 60 mV at room temperature. If the change in the meter reading is not within this range, recondition the electrode as described in the electrode instruction manual.

NOTE: Many pH/mV meters presently available can be calibrated to read concentration directly (Hach 1997, Ross 1990). It is not necessary to plot a graph when using one of these meters. Consult the meter instruction manual for this procedure and recheck calibration as recommended.

5. Transfer 25-mL sample to a beaker. Add sodium and potassium ISA and proceed as described in step 2. After obtaining the reading, use the

calibration curve to determine the sodium concentration of the sample or record the meter reading, if it has a direct concentration readout. For more information on preparing a standard curve, see the Standard Calibration Curve section in chapter 1.

Process additional samples as described in step 5.

This page intentionally blank

Chapter   **25**

Solids (Dissolved)

PURPOSE OF TEST

Dissolved solids refers to matter that is dissolved in water. Waters with high dissolved solids generally are of inferior palatability and may induce unfavorable physiological reactions, especially in transient consumers. For these reasons, a limit of 500 mg dissolved solids/L is desirable for drinking water.

LIST OF SIMPLIFIED METHODS

Several methods for determining solids are provided in Section 2540 of *Standard Methods for the Examination of Water and Wastewater*.

SIMPLIFIED PROCEDURE

Filterable Residue or Total Dissolved Solids Dried at 180°C

A well-mixed sample is filtered through a standard glass fiber filter. The filtered water is evaporated to dryness in a weighed dish and dried to constant weight at 180°C. The increase in dish weight represents the total dissolved solids. Theoretical values determined by chemical analysis may not agree with the value determined by this procedure. Approximate methods for correlating chemical analysis with dissolved solids are available (Sokoloff 1933). In general, evaporating and drying water samples at 180°C will yield values for dissolved solids closer to those obtained by adding individually determined mineral species than the dissolved solids value obtained by drying at a lower temperature.

Warnings/Cautions.

Residue from highly mineralized waters that contain significant concentrations of calcium, magnesium, chloride, or sulfate may readily take up moisture and require prolonged drying, desiccation, and rapid weighing.

Samples that contain high concentrations of bicarbonate (HCO_3^-) require careful and possibly prolonged drying at 180°C to ensure that all bicarbonate is converted to carbonate (CO_3^{2-}).

Samples that contain excessive residue may form a water-trapping crust. Limit the total sample residue to 200 mg.

Take samples in glass or plastic bottles provided so any material in suspension does not adhere to the sample bottle. Begin analysis as soon as possible because there is no practical preservation technique. Refrigerate the sample until analysis to retard possible biodegradation.

Apparatus.

- glass fiber filter disks, 4.7-cm or 2.1-cm, without organic binder[*]

- a filter holder, membrane filter funnel, or Gooch crucible adapter

- a suction flask, 500-mL capacity

- Gooch crucibles, 25-mL for 2.1-cm size filter, or 40-mL for the 4.7-cm filter size

- evaporating dishes, porcelain, 100-mL volume (Vycor or platinum dishes may be substituted)

- steam bath

- drying oven that can maintain 180°C ± 2°C

- desiccator

- analytical balance, that can weigh to 0.1 mg accuracy

- 100-mL large-tip pipet

Procedure.

1. Prepare evaporating dishes by heating clean dishes to 180°C ± 2°C for 1 hr. Cool the dishes in a desiccator and store there until needed. Weigh immediately before using on calibrated balance.

2. Prepare a glass fiber filter by placing the disk on a membrane filter apparatus or insert into the bottom of a Gooch crucible, wrinkled side up. While applying vacuum, wash the disk with three successive 20-mL volumes of distilled water. Remove all traces of water by continuing to apply vacuum after the water has passed through. Discard the washings.

3. Assemble the filtering apparatus and begin suction. Stir sample with magnetic stirrer and "transfer" 100 mL to the vacuum funnel, using a graduated cylinder. If total filterable residue is low, filter a larger volume.

4. Filter the sample through the glass fiber filter and rinse the filter with three 10-mL portions of distilled water. Continue to apply a vacuum for about 3 min after filtration is complete to remove as much water as possible.

---

[*]Glass fiber filters from Whatman grade 934H; Gelman type A/E; Millipore type AP40; E-D Scientific Specialties grade 161; or equivalent are acceptable.

5. Transfer the filtrate (100 mL, or a larger volume if the residue is low) to a weighed evaporating dish and evaporate the water sample to dryness on a steam bath or to dryness in a drying oven set to $90°C$.

6. Dry the dish that contains the evaporated sample for at least 1 hr at $180°C$ $\pm 2°C$. Cool the dish in a desiccator and weigh. Repeat the drying cycle until a constant weight is reached or until the weight loss is less than 0.5 mg on a calibrated balance.

7. Calculate total dissolved solids as follows:

$$\text{TDS, mg/L} = \frac{(A - B) \times 1,000}{C}$$

Where:

A = weight of dried residue + dish in mg
B = weight of dish in mg
C = volume of sample used in mL

This page intentionally blank

Chapter **26**

Sulfate

PURPOSE OF TEST

Sulfate is a common constituent of natural waters and is often present in drinking water, surface water, and domestic and industrial wastes. In natural waters concentrations of sulfate cover a wide range from a few milligrams per litre to several thousand milligrams per litre. High concentrations in drinking water (greater than 500 mg/L) may cause diarrhea and dehydration. This is of special concern for infants.

LIST OF SIMPLIFIED METHODS

The simplest methods available to test for sulfate in water are the turbidimetric or spectrophotometric method (*Standard Methods for the Examination of Water and Wastewater,* Section 4500–SO_4^{2-} E.), applicable for a range of 1–40 mg/L sulfate. Gravimetric analysis (*Standard Methods* 4500 SO_4^{2-} C. and D.) is applicable for concentrations greater than 10 mg/L. However, even though they require no instrumentation, gravimetric methods are time consuming and require very precise analytical skill. Ion chromatographic (*Standard Methods* 4110 or USEPA Method 300.0) methods are very accurate and have a detection limit of approximately 0.1 mg/L sulfate.

Refer to *Standard Methods* Section 4500–SO_4^{2-} Sulfate. The US Environmental Protection Agency (USEPA) recommends USEPA Method 300.0 or 375.2 or *Standard Methods* 4500 SO_4^{2-}, C., D., and E. for monitoring purposes.

SIMPLIFIED PROCEDURE

Turbidimetric/Spectrophotometric Method (*Standard Methods* 4500–SO_4^{2-} E.)

Sulfate ion is precipitated as barium sulfate ($BaSO_4$) by adding barium chloride ($BaCl_2$) to the sample. The turbidity of the barium sulfate suspension is then measured using a turbidimeter or a spectrophotometer. The concentration of sulfate is determined by comparing the results to a standard curve.

Warnings/Cautions.

Perform the test with samples and standards at room temperature.
Keep the mixing speed and time constant for all samples and standards.
Store samples at 4°C.
Color or suspended solids will interfere. Remove suspended solids by filtering the sample through a glass fiber filter.
Silica at concentrations higher than 500 mg/L will interfere.

Apparatus.

- a turbidimeter (nephelometer), or spectrophotometer at a wavelength of 420 nm with a cell length of 2.5 to 10 cm, or a filter photometer with a violet filter with a maximum wavelength near 420 nm and a cell length of 2.5 to 10 nm

- a magnetic stirrer with adjustable speeds

- a timer

- a measuring spoon, capacity 0.2 to 0.3 mL

- a stir bar

- one 250-mL Erlenmeyer flask for each standard and sample

- five 100-mL volumetric flasks

- five volumetric pipets of sizes 5, 10, and 20 mL

Reagents.

Conditioning Reagent A. Dissolve 30 g magnesium chloride ($MgCl_2 \cdot 6H_2O$), 5 g sodium acetate ($CH_3COONa \cdot 3H_2O$), 1.0 g potassium nitrate (KNO_3), and 20 mL 99 percent acetic acid (CH_3COOH) into 500 mL distilled water, and dilute to 1 L.

Conditioning Reagent B. This reagent is used if the sulfate concentration is less than 10 mg/L. Prepare the reagent with all the chemicals listed for conditioning reagent A, and add 0.111 g sodium sulfate (Na_2SO_4).

Barium chloride crystals, 20 to 30 mesh. Prepackaged reagents can be used in place of the conditioning reagent and the barium chloride crystals ($BaCl_2$). Follow the manufacturer's instructions.

Sulfate standard solution. Dilute 0.1479 g anhydrous sodium sulfate (Na_2SO_4) in distilled water and dilute to 1 L. The concentration of this standard is 100 mg/L sulfate. Store the standard solution at 4°C. Alternatively, purchase stock sulfate standard.

All reagents are commerically available.

Procedure.

1. Using a volumetric pipet, measure 100-mL sample into a 250-mL Erlenmeyer flask. If the sulfate concentration is higher than 40 mg/L, dilute a portion of the sample to 100 mL in a volumetric flask.

2. Add 20 mL conditioning reagent.

3. Insert a stir bar and begin stirring on a magnetic stirrer. Increase the stirring speed so it is just below the speed at which splashing occurs. Mark the speed setting and use this speed for all analyses.

4. While stirring, add one measuring spoonful of barium chloride crystals and begin timing immediately.

5. Stir for 60 ± 2 sec at a constant speed.

6. Pour the solution into the measuring cell and place cell into the turbidimeter or spectrophotometer.

7. Record the reading after the sample has been allowed to stand for 5 ± 0.5 min.

8. Prepare a calibration curve by diluting a portion of the 100 mg/L stock standard sulfate solution. Use a volumetric pipet to measure the stock standard into a 100-mL volumetric flask (Table 26-1). Dilute to 100 mL using distilled water. Analyze the standard solutions using this procedure.

9. Plot the calibration curve on graph paper, with concentration on the x axis and either the turbidimeter readings (NTU) or the percent transmittance from the spectrophotometer on the y axis.

10. Use the graph and the sample readings to determine the sulfate concentration. Correct the concentration for any dilutions made for each sample. For more information on preparing a standard curve, see the Standard Calibration Curves section in chapter 1.

Table 26-1   Dilutions to prepare sulfate solutions for calibration curve

Volume of Stock Standard mL	Concentration of Working Standard mg/L
5	5
10	10
20	20
30	30
40	40

This page intentionally blank

Chapter **27**

Taste and Odor

PURPOSE OF TEST

One of the primary objectives of a water treatment facility is to produce a finished water that is free of objectionable taste and odor. This parameter is closely monitored by the utilities' customers and can quickly give rise to numerous complaints when there is a problem. The threshold odor test is a valuable test for evaluating taste and odor problems.

Unpleasant tastes and odors in water can come from several sources, including algae blooms, bacteria, actinomycetes, decaying vegetation, and industrial waste. Tastes and odors can also result from pipe corrosion. Corrective action varies depending on the source of the problem and the capabilities of the treatment facilities. Treatment measures include aeration, chlorine dioxide, activated carbon (both powdered and granular), chlorine, copper sulfate (for algae control), potassium permanganate, and flushing water mains. Other techniques may also be applicable.

LIST OF SIMPLIFIED METHODS

The threshold odor test is the basis by which odor intensity is measured. In this test, the odor-bearing water sample is diluted in incremental stages with odor-free water until the water has a barely perceptible odor. Refer to *Standard Methods for the Examination of Water and Wastewater*, Section 2150 Odor and Section 2170 Flavor Profile Analysis. For additional information, see *Flavor Odor and Profile Analysis* report (reference in appendix A).

SIMPLIFIED PROCEDURE

Threshold Odor Test

Warnings/Cautions.

This test relies on the sense of smell, but the same procedure can be performed using taste. If taste testing is performed, only waters known to be safe for drinking may be tested, and the water must not be dechlorinated.

Apparatus.

- glass-stoppered sample bottles with TFE-lined tops. Do not use plastic containers. All glassware must be odor free and rinsed with 1:1 HCL then odor-free water before use.

- six 500-mL Erlenmeyer flasks, five with ground-glass stoppers (ST-32), one without ground-glass stopper

- thermometers, $20°-110°$ C

- a 100-mL graduated cylinder

- a 50-mL graduated cylinder

- a 25-mL graduated cylinder

- a 10-mL measuring pipet

- a large electric hot plate, stove top or waterbath

- odor-free water generator, granular activated carbon filter

- two 2- or 4-L flasks to collect and warm odor-free water

- a 500-mL Erlenmeyer flask, with odor-free water

Reagents.

Odor-free water. Pass ordinary drinking water through a granular activated carbon filter at a rate of approximately 0.1 L per min. When first starting the generator, flush to remove all carbon fines before using the odor-free water. Depending on the number of water samples to be tested, generate 2–3 L of odor-free water. Odor-free water should be generated fresh each day because odors in the laboratory may contaminate the water, or use de-ionized water.

Dechlorination reagent. Dissolve 3.5 g sodium thiosulfate ($Na_2S_2O_3 \cdot 5H_2O$) in water and dilute to 1L. Or dissolve 0.93 g sodium arsenite ($NaAsO_2$) in water and dilute to 1 L. **Caution: Toxic—do not ingest.** Review the Material Safety Data Sheet carefully before using or preparing sodium arsenite solution.

Dechlorination agent should be added to the test blank also.

No preservation is required as long as the chlorine residual is neutralized before testing.

Procedure.

1. Place 200 mL odor-free water into a 500-mL Erlenmeyer glass-stoppered flask as the odor-free blank.

2. Pipet or pour various amounts of the sample (test immediately after collection) into a series of 500-mL Erlenmeyer flasks (such as 100-, 50-, 25-, 12-, and 2.8-mL). Then add odor-free water so the final volume in the flask is 200 mL. Also prepare a flask with 200 mL undiluted sample.

3. Mix stoppered samples by swirling the flask while warming them on the hot plate. Heat all flasks to 60°C, checking the temperature regularly on a separate "dummy" flask. Considerable time can be saved by preheating the odor-free water to 60°–65°C.

4. After all the samples and the odor-free blank have reached 60°C, shake the sample to release vapors, then remove the stoppers and sniff the vapors. Smell and compare each sample against the odor-free blank, beginning with the most diluted sample. Continue until all dilutions have been tested.

5. Record which flasks contain an odor and which do not. The sample that has the least detectable odor is the end point. The actual threshold odor (TO) number is determined by the number of dilutions needed to bring the sample to a point where the odor is barely detectable. The number of persons selected for the analysis may vary, but the panel should consist of at least 3 people or an odd number in order to break "tie" responses. These people should be free of allergies and perfumes. The test should be conducted in an odor free environment.

6. Calculations for TO

$$\text{TO} = \frac{A + B}{A} \qquad (27\text{-}1)$$

Where:

A = volume of sample (mL)
B = volume of odor-free water used for dilution (mL)

Example: The volume of sample plus odor-free water (Table 27-1) must always equal 200 mL.

The odor is no longer detectable at 70 mL of sample; 80 mL of sample is the lowest detectable point; therefore, the TO of the sample should be based on 80 mL of sample.

Based on the previous example, the TO calculation would be as follows:

$$\text{TO} = \frac{80 + 120}{80} = 2.5$$

The TO of the example is 2.5.

Table 27-1   Preparing samples for threshold odor analyses

Sample Volume		Odor-Free Water mL		Test Odor
110	+	90	+	(positive for odor)
100	+	100	+	
90	+	110	+	
80	+	120	+	
70	+	130	−	(no odor detected)
60	+	140	−	

Chapter **28**

Temperature

PURPOSE OF TEST

Accurate temperature readings are important in many treatment processes and laboratory determinations. For example, temperature is a factor in certain algal blooms, in the degree of dissolved-oxygen saturation, and in carbon dioxide concentration.

LIST OF SIMPLIFIED METHODS

Standard Methods for the Examination of Water and Wastewater, Section 2550, contains information on several methods of temperature measurement.

SIMPLIFIED PROCEDURE

Temperature

Warnings/Cautions.

For best results, take the temperature at the same sampling point each time a sample is taken. Immerse the thermometer in the flowing stream, or in a large container filled with the sample and held until the mercury level is steady. Read the temperature before removing the thermometer from the sample.

Mercury is a poison, so mercury thermometers should be permanently installed only in a water pipe that leads to a drain, rather than one that leads into the distribution system. If the thermometer breaks and is installed to a drain, it cannot cause a serious spill of mercury into the drinking water supply.

Apparatus.

- A mercury-filled Celsius thermometer with a range of $0°-100°$ C can be used for most purposes. At a minimum, the scale should be subdivided into $0.1°$ C.

Thermometers are calibrated for total or partial immersion. Total-immersion thermometers must be completely immersed in water to yield the correct temperature. Partial-immersion thermometers must be immersed only to the depth of the etched circle that appears around the stem just below the scale level. For best results, the accuracy of the thermometer in routine use should be checked against a precision thermometer certified by the National Institute of Standards and Technology.

- An all-metal dial thermometer may also be used but must be calibrated.

- An electronic thermometer with a thermistor and digital readout must also be calibrated. Follow the manufacturer's instructions for use and calibration.

Procedure.

1. Immerse the thermometer in the sample to the proper depth for a correct reading.

2. Record the temperature to the nearest fraction of a degree Celsius before removing. If thermometer readings must be converted between Celsius and Fahrenheit, refer to Table 28-1.

Table 28-1   Temperature conversions

°F	°C	°F	°C	°F	°C	°F	°C
30	−1.1	49	9.4	68	20.0	87	30.6
31	−0.56	50	10.0	69	20.6	88	31.1
32	0.0	51	10.6	70	21.1	89	31.7
33	0.56	52	11.1	71	21.7	90	32.2
34	1.1	53	11.7	72	22.2	91	32.8
35	1.7	54	12.2	73	22.8	92	33.3
36	2.2	55	12.8	74	23.3	93	33.9
37	2.8	56	13.3	75	23.9	94	34.4
38	3.3	57	13.9	76	24.4	95	35.0
39	3.9	58	14.4	77	25.0	96	35.6
40	4.4	59	15.0	78	25.6	97	36.1
41	5.0	60	15.6	79	26.1	98	36.7
42	5.6	61	16.1	80	26.7	99	37.2
43	6.1	62	16.7	81	27.2	100	37.8
44	6.7	63	17.2	82	27.8	101	38.3
45	7.2	64	17.8	83	28.3	102	38.9
46	7.8	65	18.3	84	28.9	103	39.4
47	8.3	66	18.9	85	29.4	104	40.0
48	8.9	67	19.4	86	30.0		

Chapter **29**

Turbidity

PURPOSE OF TEST

Turbidity measurement results are used to control the amount of coagulant and other chemical aids that produce a water of the desired clarity. Public water supplies are coagulated and filtered to reduce suspended particles to an acceptable level.

> **NOTE: Turbidity is now used by the US Environmental Protection Agency as an indicator of the possible presence of biological organisms. Turbidity measurements of finished drinking waters normally are in the 0 to 1 ntu range.**

Turbidity is composed of insoluble particles of clay, silt, mineral matter, organic debris, plankton, and other microscopic organisms that impede or scatter the passage of light through water. Turbidity is an expression of the optical property that causes light to be scattered or absorbed rather than transmitted in straight lines through the sample. Correlation of turbidity with the weight concentration of suspended matter is difficult because of the size, shape, and refractive index of particles (Hach 1972).

The method is based on a comparison of the intensity of light scattered by the sample under defined conditions with the intensity of light scattered by a standard reference suspension. The greater the scattering of the light, the higher the turbidity. Readings (in ntu) are made in a nephelometer designed according to specifications outlined in the Specific Purpose Laboratory Equipment section in chapter 1. A standard suspension of formazin, prepared under closely defined conditions, is used to calibrate the instrument (see *Standard Methods for the Examination of Water and Wastewater,* Section 2130 B. Nephelometric Method).

LIST OF SIMPLIFIED METHODS

The nephelometric procedure is the only procedure applicable to finished drinking waters.

SIMPLIFIED PROCEDURE _____

Nephelometric Turbidity

Warnings/Cautions.

Hydrazine sulfate $[(NH_2)_2 \cdot H_2SO_4]$ used to prepare the formazin solution suspension for calibration of the nephelometer is a carcinogen. Once the polymerization is complete, the formazin suspension (preparation is described under Reagents), is not considered to be a carcinogen. However, laboratories that wish to eliminate the hazard of hydrazine sulfate exposure may purchase the stock 4,000 ntu formazin standard suspension from commercial vendors and eliminate the need to purchase, store, and handle hydrazine sulfate.

Samples should be free of debris and rapidly settling coarse sediments. Dirty glassware, air bubbles, and vibrations that disturb the surface of the sample will give false results.

Apparatus.

- A turbidimeter that consists of a nephelometer with a light source for illuminating the sample and one or more photoelectric detectors with a readout device to indicate the intensity of light scattered at right angles to the path of the incident light. The turbidimeter should be designed so minimal light reaches the detector in the absence of turbidity and should be free from significant drift after a short warmup period.

  Instrument sensitivity should permit detecting turbidity differences of 0.02 ntu or less in waters with a turbidity less than 1 ntu. The range of the instrument should be 0–10 units. Several ranges may be necessary to achieve this coverage.

  The following design criteria should be observed to minimize differences in measured values: (1) Light source is a tungsten lamp operated at a color temperature of 2,200 to 3,000° K; (2) total distance traversed by the incident light and scattered light in the sample should not exceed 10 cm; (3) the detector should be centered at 90° to the incident light path. The detector should have a spectral peak between 400 and 600 nm.

- Sample tubes to be used with the instrument must be of clear, colorless glass. The tubes must be kept scrupulously clean, both inside and out, and discarded when they become scratched or etched. The tubes must not be touched where light strikes them.

Reagents.

- *Turbidity–free water.* Pass distilled water through a 0.2-μm filter. Rinse the collecting flask at least twice with filtered water. Discard the initial volume of water. Commercial water that meets these requirements may be purchased.

- *Stock turbidity solution.* Formazin polymer is used as the turbidity reference suspension for water because it is more reproducible than standards previously used for turbidity. An acceptable alternative primary standard is the AMCO AEPA-1 polymer beads. However, the AMCO standards should not be used for low-range turbidity (less than 0.2 ntu) calibrations of ratio turbidimeters.

1. Solution 1. Dissolve 1.00 g hydrazine sulfate in distilled water and dilute to 100 mL in a volumetric flask.

2. Solution 2. Dissolve 10.00 g hexamethylenetetramine in distilled water and dilute to 100 mL in a volumetric flask.

3. In a 100 mL flask, mix 5.0 mL solution 1 with 5.0 mL solution 2. Allow to stand 24 hr at 25° C, then dilute to the mark and mix well. This is the stock formazin solution.

Standard formazin turbidity suspension. Dilute 10.00 mL stock turbidity suspension to 1,000 mL with turbidity-free water. The turbidity of this water is defined as 40 units. Dilute portions of the standard turbidity suspension as required. Prepare these daily.

NOTE: **To avoid contact with the hydrazine sulfate, a stock 4,000 ntu standard suspension can be purchased from commercial vendors.**

Procedure.

Follow the manufacturer's operating instructions to calibrate the turbidimeter. Prepare and measure standards on the turbidimeter that cover the range of samples to be tested. At least one standard should be run in each range used. Based on determination of a standard, make certain the instrument is stable in all ranges used. High turbidities determined by direct measurements are likely to differ appreciably from those determined by the dilution technique.

Samples may be held at 4° C for a maximum of 48 hr, but the measurement should be performed as soon as possible.

To measure turbidities less than 40 ntu, shake the sample thoroughly to disperse the solids. Wait until all air bubbles disappear, then pour the sample into a turbidimeter tube. Read the turbidity directly from the calibration curve.

To measure turbidities greater than 40 ntu dilute the sample with sufficient equal volumes of turbidity-free water to achieve a turbidity of 30–40 ntu. Calculate turbidity of the original sample from the turbidity of the diluted sample with the dilution factor.

Example: One volume of the original sample and three volumes of dilution water equal 1+3 times the determined value.

This page intentionally blank

Chapter **30**

Bacteriological Examination

PURPOSE OF TEST

Coliforms are common in the environment and are generally not harmful. Although not disease producers themselves, coliforms are often associated with pathogenic organisms. As such, they provide a good index of the degree of bacteriological safety of a water. For that reason, the US Environmental Protection Agency (USEPA) has determined that the presence of fecal coliforms or *Escherichia coli (E. coli)* in drinking water is a serious health concern.

In polluted water, coliform bacteria are found in densities roughly proportional to the degree of fecal pollution. Their presence in drinking water is generally caused by a problem with water treatment or distribution systems. The presence of coliforms indicates the water may be contaminated with organisms that can cause disease.

Disease symptoms may include diarrhea, cramps, nausea, and possibly jaundice, associated headaches, and fatigue. However, these symptoms may be caused by many other factors.

USEPA has set an enforceable drinking water standard for total coliform to reduce the risk of these health effects. According to this standard, no more than 5.0 percent of the samples collected during a month can contain these bacteria. Water systems that collect less than 40 samples per month and have one total coliform-positive sample per month are not violating the standard.

In June 1990, USEPA published in the *Federal Register* the final rule establishing a maximum contaminant level (MCL) for total coliform, fecal coliform, and *E. coli*. The rule regulates monitoring and analytical requirements. The MCL is based on the presence or absence of total coliforms in a sample, rather than on coliform density.

USEPA has identified the following as the best available technology (BAT), and treatment techniques for achieving compliance with the MCL:

1. protection of water wells by appropriate placement and construction

2. maintenance of a disinfectant throughout the distribution system

3. proper maintenance of the distribution system, including appropriate pipe replacement and repair procedures, main flushing programs, proper operation and maintenance of storage tanks and reservoirs, and continual maintenance of positive water pressure in all parts of the distribution system

4. filtration or disinfection of surface water using chlorine, chlorine dioxide, or ozone

The Total Coliform Rule requires that a utility develop and implement a written sample siting plan to monitor for coliforms. A well-designed siting plan provides early evidence of changing water quality characteristics and bacteriological problems. The siting plan can track changes in these key bacterial populations as well as loss of residual disinfectant in the distribution system.

Routine samples for bacteriological examination should be collected from representative points in the distribution system. The number of samples examined per month is based on the total population served by the water supply. A systematic, routine plan covering all major areas of the distribution system is critical, and the plan will vary from system to system. Areas of concern include slow flow areas, dead ends, first customer locations, all reservoirs and tank outflows, water supply in open reservoirs, and all service areas of the pipe network. Sampling frequency depends on operating conditions.

The Surface Water Treatment Rule (SWTR) includes regulations for heterotrophic bacteria and other pathogenic organisms that are removed by treatment techniques. The SWTR applies to all public water systems that use surface water sources or groundwater sources under the direct influence of surface water.

LIST OF SIMPLIFIED METHODS

Refer to *Standard Methods for the Examination of Water and Wastewater*, Sections 9221–9225, for an in-depth description of this methodology.

Methods covered in this chapter include

- Total Coliform

    — ONPG-MUG, MMO-MUG, or Colilert™, *Standard Methods*, Sections 9223 A. and B., USEPA approved

    — Colisure™, USEPA approved

    — Presence–absence method, *Standard Methods*, Section 9221 D., USEPA approved

Methods mentioned include membrane filtration, *Standard Methods*, Sections 9222 A., B., and C., USEPA approved; and multiple tube fermentation, *Standard Methods*, Sections 9221 A., B., and C., USEPA approved.

- Fecal Coliform

    — EC medium confirmation, USEPA approved

- Heterotrophic plate count, *Standard Method* 9215 B., USEPA approved

- *E. coli*

— EC medium + MUG, USEPA approved

— Nutrient agar + MUG, USEPA approved

For more information on the fundamental theories and concepts of bacteriological testing, see *An Operator's Guide to Bacteriological Testing.*

SIMPLIFIED PROCEDURES

Sample Collection Information

Each public water system must sample according to a written sample siting plan that is subject to review and revision. The number of samples taken each month is based on population size. A portion of the complete list is provided in Table 30-1. To see the complete list, contact your local health department or refer to the *Federal Register.*

If a routine sample is total coliform positive, the public water system must collect a set of repeat samples within 24 hr of being notified of the positive result.

A system that collects more than one routine sample per month must collect at least three repeat samples for each total coliform-positive sample found. A system that collects one routine sample per month or fewer must collect no fewer than four repeat samples for each total coliform positive sample found. The system must collect at least one repeat sample from the sampling tap where the original total coliform-positive sample was taken, at least one repeat sample at a tap within five service connections upstream, and at least one repeat sample at a tap within five service connections downstream of the original sampling site.

The system must collect all repeat samples on the same day. If one or more repeat samples in the set are total coliform positive, the public water system must collect an additional set of repeat samples in the manner specified previously. This procedure must be repeated until either total coliforms are not detected in one complete set of repeat samples or the system determines that the MCL for total coliforms has been exceeded and notifies the state.

If a system that collects fewer than five routine samples per month has one or more total coliform-positive samples, it must collect at least five routine samples during the next month the system provides water to the public.

Results of all routine and repeat samples not invalidated by the state must be included in determining compliance with the MCL for total coliforms.

Table 30-1   Selected bacteriological samples based on population size

Number of Population Served			Routine Samples/Month
25	to	1,000	1
1,001	to	2,500	2
2,501	to	3,300	3
3,301	to	4,100	4
4,101	to	4,900	5
25,001	to	33,000	30
130,001	to	220,000	120
3,960,001	or	more	480

Sample Collection Procedures

Collect all samples in bottles or bags that can hold 120 mL. Containers must be sterilized and pre-dosed with sodium thiosulfate ($Na_2S_2O_3 \cdot 5H_2O$). Sample bags are purchased sterile, with a pellet of sodium thiosulfate added to them. Sample bottles must contain 0.1 mL of 10 percent sodium thiosulfate solution (0.010 grams per 100 mL distilled water) before sterilization in the autoclave at 121°C for 15 min at 15 lb pressure.

Keep the sample container closed until the moment of sample collection. Avoid collecting samples on rainy or windy days because of the increased possibility of contamination. When collecting the sample, do not lay the bottle lid down or hold it face up because of the possibility of contamination. Do not rinse the container. Fill the sample container to the line on the bags or about four-fifths full for the bottles. When bottles are received, mark them at the 100 mL mark to ensure the correct amount of sample is taken.

When collecting the sample, observe the following procedures:

1. If the sample is from a cold water tap, remove any screen or other attachment. Let the water run in a hard stream for 5 min. Decrease the flow and fill the sample bottle. Immediately record the site, date, time, and sampler's initials. Measure and record the disinfectant residual.

2. If the sample must be taken from a mixing faucet (not recommended), remove any screen or other attachment and let the hot water run for 2 to 3 min. Run the cold water for 2 min and decrease the flow to a steady stream to take the sample. Record the site, date, time, and sampler's initials. Note that the sample was from a mixing faucet. Measure and record the disinfectant residual.

3. When sampling a source of supply such as a lake, river, or impoundment, immerse the bottle, open mouth held vertically to the water, a few inches below the surface. This helps to avoid surface scum and debris. Turn the mouth of the bottle into the current or create a current by slowly moving a hand through the water. Pour a small amount of water from the full bottle and cap. Record the site, date, time, and sampler's initials.

Refrigerate or ice the sample. If ice is used, it should be in a container such as blue ice—not wet. Return it to a certified laboratory as soon as possible. Environmental samples must be analyzed within 6 hours of collection. Chlorinated samples must be analyzed within 30 hr of collection.

Table 30-2 indicates methods used for microbiological testing to identify total coliform, fecal coliform, and *E. coli*.

Total Coliform

Federal and state agencies require the analysis of total coliform in drinking water samples to provide information about the bacteriological quality of the water. Several analytical methods are approved for this analysis. Some are quantitative and provide an approximate count of the number of colonies in the sample. Examples are the membrane filtration and multiple tube methods. Other techniques are qualitative and provide only a *presence* or *absence* response. The easiest analyses to use are qualitative. For any of these analyses, pre-made reagents greatly facilitate the test performance.

ONPG-MUG, MMO-MUG, OR COLILERT™
(*Standard Methods* 9223 B.)

Colilert™ is a proprietary system approved by USEPA for the analysis of total coliform and *E. coli* in drinking water. In the MMO-MUG technique, a water sample is added to a tube inoculated with the dry medium, which is selective and suppresses the growth of noncoliform bacteria. Test results are determined by a visual inspection of the tube under standard and ultraviolet (UV) light during or at the end of a 24-hr incubation period. If coliform bacteria are present, the medium is metabolized and colored compounds are released. Yellow indicates the presence of total coliform. Yellow with fluorescence under UV light indicates the presence of *E. coli*. No color development indicates a negative sample result. The MMO-MUG technique is a confirmed test after 24 hr. If color development is uncertain, an additional 4 hr of incubation may be needed. In situations with exceptionally high heterotrophic plate counts (more than 500 colony forming units per 1 mL), false positive results may be produced.

Several variations are possible when analyzing drinking water with the MMO-MUG technique. The most probable number (MPN) format or the presence–absence (P–A) format may be used. Sterile, disposable tubes with the medium predispensed may be used, or a measured dose of medium can be added to the laboratory's own glassware, which must be sterilized in the same manner as the glassware used in the standard MPN or membrane filtration (MF) coliform procedures.

Table 30-2   Microbiological techniques

Total Coliform	Format	Confirmation	Fecal Coliform	E. coli
Membrane filter	1 × 100 mL	LT and BGB (Coliform)	EC Broth with filter transfer	Nutrient agar + MUG with filter transfer
Multiple tube fermentation	1 × 100 mL 5 × 20 mL 10 × 10 mL	BGB	EC Broth with swab inoc. EC Broth with loop inoc.	EC Broth + MUG with swab inoc. EC Broth + MUG with loop inoc.
P–A	1 × 100 mL	BGB	EC Broth with loop inoc.	EC Broth + MUG with loop inoc.
ONPG-MUG (MMO-MUG) (Colilert™)	1 × 100 mL 5 × 20 mL 10 × 10 mL 100 × 1 mL	None		UV light EC Broth + MUG with pipet inoc.
Colisure™	1 × 100 mL	None		UV light
P–A + MUG*	1 × 100 mL	BGB*		UV light*
API Strips*				
Enterotubes*				

Where:

*	=	Not USEPA approved
BGB	=	Brilliant green lactose bile broth
P–A	=	Presence–Absence
LT	=	Lauryl tryptose broth
EC Broth	=	*Standard Methods* EC fecal broth
ONPG-MUG	=	Ortho-nitrophenyl-β-D-galactopyranoside + 4-methylumbelliferyl-β-D-glucuronide
Inoc.	=	Inoculation

Alternate Procedure 1.

1. For the MPN format, use a sterile pipet to transfer 20- or 10-mL portions of a water sample into a series of 5 or 10 tubes, respectively, with the MMO-MUG media. Use a new pipet for each water sample. Uncap the tubes just before adding the water sample and recap immediately after adding the sample.

2. After adding the water sample to the entire series of tubes, shake the planted tubes vigorously and repeatedly invert them to dissolve the media. Some media particles may remain undissolved. Another option allows the inoculation of 100 vials in a 10 × 10 tray with 1 mL sample each. This option is suitable for a low density of coliforms.

3. Place planted tubes in an incubator within 30 min after inoculation. The incubation temperature should be between 35° and 37° C, with 35° ± 0.5° C preferable.

4. After 24 hr of incubation, inspect the samples for color change under normal lighting conditions. As indicated earlier, no color development is a negative result for total coliforms and *E. coli*. Yellow is a positive result for total coliform. If color development is uncertain, use a color comparator. Any color development darker than the comparator is positive. Color development lighter than the comparator should be incubated for an additional 4 hr, but not more than a total of 28 hr. The additional 4-hr incubation time increases color development if coliforms are present. After 28 hr of incubation, only tubes without color change are considered valid.

5. Examine all tubes with color development under UV light in a dark environment for fluorescence. Any yellow tubes with fluorescence are positive for *E. coli*.

6. When examining the tubes, record the number of tubes that show yellow with or without fluorescence. The total coliform MPN is determined by comparing the number of positive tubes with the appropriate MPN Index Table, such as Table 9222:II from *Standard Methods*. The *E. coli* MPN is determined in the same manner using the number of tubes that show fluorescence. No other verification or confirmation tests are required.

Alternate Procedure 2.

1. In the P–A format, a 100-mL sample is added to a sterile vessel adequately dosed with the MMO-MUG medium.

2. Seal the vessel immediately, shake, and invert several times. A few particles of the medium may not completely dissolve.

3. Incubate the vessel for 24 to 28 hr at 35° to 37° C with 35° ± 0.5° C preferable. At the end of the incubation period, check the vessel for color development.

4. If yellow is present, examine the vessel in a dark environment under UV light for fluorescence. Yellow is a positive result for total coliform. If color development is uncertain, use a color comparator. Any color development darker than the comparator is positive, and color development less than the comparator should be incubated for an additional 4 hr but not more

than a total of 28 hr. The additional 4-hr incubation time increases color development if coliforms are present. After 28 hr of incubation, only tubes without color change are considered valid.

5. Examine all vessels with color development under a UV light in a dark environment for fluorescence. Any yellow vessels with fluorescence are positive for *E. coli*.

6. Report results as absent or less than one (<1) total coliform per 100 mL of sample if there is no color development. With color development without fluorescence, report results positive for total coliform. With both color development and fluorescence, report the results as positive for *E. coli*.

For either alternative, discard the used media and any disposable glassware properly. One possible procedure requires that caps on the used reagent tubes or vessels be loosened and both be placed in an autoclave for 20 min at 121°C. After autoclaving, flush used media down the drain and discard any disposable glassware. Alternatively, add a sodium hypochlorite solution, such as bleach, to the tubes or vessels and allow them to sit for at least 20 min before discarding.

Colisure™ Method

The Colisure™ test is a proprietary system approved by USEPA for the analysis of total coliform and *E. coli* in drinking water. The procedure requires 100 mL of sample to be added to a vial that contains the Colisure™ medium. Incubate the sample for 24 hr at 35° ± 0.5°C. If the color changes from yellow to purple (or magenta), the sample is positive for total coliform. Samples that remain yellow are negative for total coliform. Examine all positive coliform samples under UV light for fluorescence. If there are any questions regarding color change or fluorescence, incubate the sample for an additional 24 hr. Report the results as present or absent.

Procedure.

1. Add 100 mL sample to a sterile vessel adequately dosed with the Colisure™ medium.

2. Seal the vessel immediately, shake, and invert several times. A few particles of the medium may not completely dissolve. The initial color is yellow.

3. Incubate the vessel for 24 to 28 hr at 35° ± 0.5°C. At the end of the incubation period, check the vessel for development of a magenta purple to a dirty purple color.

4. If purple is present, examine the vessel in a dark environment under UV light for fluorescence. Purple is a positive result for total coliform. If color development is uncertain, incubate the sample for an additional 4 hr and reexamine. If a color change is still uncertain, incubate the sample for an additional 20 hr but not more than a total of 48 hr. The additional incubation time increases color development if coliforms are present. After 48 hr of incubation, only vessels with a definite color change are considered positive for total coliform.

5. Examine all vessels with purple under a UV light in a dark environment for fluorescence. Any purple vessels with fluorescence are positive for *E. coli*.

6. Report the results as absent or less than one (<1) total coliform per 100 mL of sample if there is no color change to purple. Purple without fluorescence indicates positive for total coliform. With both a purple color development and fluorescence, the results are positive for *E. coli*.

7. After the test has been completed, either autoclave or disinfect used media and glassware with a high concentration of sodium hypochlorite (NaOCl). Place vials in the autoclave for 20 min at 121°C. Alternatively, allow the sodium hypochlorite to be in contact with the positive media for at least 20 min before pouring down the drain.

Membrane Filtration Method
(*Standard Methods*, Section 9222 A., B., and C.)

The membrane filtration (MF) method for analysis of total coliform bacteria is a long, complex, quantitative method that requires 24 hr for a presumptive colony count. It requires training and experience to determine which colonies to count as total coliform on the filter and which not to count. For an in-depth discussion of this method, refer to *Standard Methods*.

Presence–Absence Method
(*Standard Methods*, Section 9221 D.)

The presence–absence (P–A) method is a modification of the multiple tube fermentation (MTF) technique approved by USEPA for detecting total coliform in drinking water. The P–A test uses a single inoculation bottle instead of the series of tubes required in the MTF procedure. The P–A method employs a mixture of lactose broth, lauryl tryptose broth, and bromcresol purple as its medium, as described in *Standard Methods*.

For 100-mL samples, prepare the media formulation in triple strength. Add 50 mL medium to a 250-mL milk dilution bottle or other suitable vessel. A fermentation tube in the medium is optional. Autoclave the vessel with medium for 12 min at 121°C, with the total time in the autoclave not to exceed 30 min. The final pH after sterilization should be 6.8 ± 0.2. The test procedure is highly subject to interference from high concentrations of heterotrophic bacteria (noncoliforms).

Procedure.

1. Mix medium and the entire 100-mL sample in a single vessel by inverting the vessel four to five times.

2. Incubate the inoculated vessel at $35° \pm 0.5°C$ and inspect after 24 and 48 hr.

3. The P–A test relies on either gas production (similar to the MTF test if a Durham tube is used), or acid production caused by media fermentation. In acid conditions, the medium has a yellow color because of the color indicator, bromcresol purple. Either condition, gas production or a color change, alone constitutes a positive sample. As with the MTF test, the P–A test must be confirmed in brilliant green lactose bile (BGLB) broth. For SDWA regulatory compliance monitoring samples, confirmation for the possible presence of fecal coliforms or *E. coli* may be required.

Fecal Coliform/Confirmation Techniques

Colilert™ and Colisure™ are USEPA-approved methods for detecting *E. coli*. The other two methods discussed are for confirmation following membrane filtration.

Reagents.

EC Medium. May be purchased ready to use in tubes or rehydrate 37 g dry medium with 1 L distilled water. Heat gently to dissolve the medium completely. Add a sufficient volume of medium so that even after sterilization the gas tube is slightly covered. Dispense 30–50 mL volumes into fermentation tubes (150 by 25 mm containing 75 by 10 mm inverted gas tubes). Autoclave for 15 min at 15 lb pressure and 121°C. Be sure to allow the temperature to drop below 75°C before opening the autoclave to avoid trapping air bubbles in the inverted tubes.

EC medium + MUG broth. Can be purchased ready to use in tubes.

Procedure.

Place a loop loaded with laurel tryptose broth and sample into a tube of EC medium. Gently swirl the tube to mix and incubate in a waterbath at 44.3°C to 44.7°C for 22 to 26 hr. Gas production of any amount in the inner fermentation tube of the EC medium indicates a positive fecal coliform test.

The same procedure is used with the EC + MUG medium. The production of gas indicates fecal coliform, and a fluorescence using a short-band UV wavelength indicates positive for *E. coli*.

Heterotrophic Plate Count/Pour Plate Method

Heterotrophic plate count (HPC) is a procedure to estimate the number of live bacteria in water and measure changes during water treatment and in water distribution systems. Bacteria cells may occur in pairs, chains, clumps, or single cells, each of which may develop into a single colony; all are included in the term *colony-forming units* (CFUs). The pour plate and membrane filter methods are described for convenience. However, sample volume and turbidity must be considered before selecting a method. Only the pour plate method (*Standard Methods* 9215 B.) is USEPA approved for SDWA compliance monitoring.

The pour plate method is simple to perform and can accommodate sample volumes or diluted samples ranging from 0.1 to 1.0 mL.

Apparatus.

- Erlenmeyer flasks
- a 100-mL graduated cylinder
- a hot plate
- 250-mL milk dilution bottles
- sterile 1-, 5-, 10- or 25-mL pipets
- sterile petri dishes (plastic or glass)
- a 44°–46°C water bath
- a 35°–37°C incubator

- an autoclave 121°C

- a pH meter

- a Quebec dark field colony counter

Reagents.

Plate count agar (tryptone glucose yeast agar)
Phosphate buffer solution

Procedure.

Selecting Dilutions. Select the dilution(s) so the total number of colonies on a plate will be between 30 and 300. For most drinking water samples, plates suitable for counting will be obtained by plating 1 mL and 0.1 mL undiluted sample and 1 mL of a 10-2 dilution.

Measuring sample and dilutions.

1. Use a sterile pipet to make transfers from each dilution bottle.

2. Use a separate pipet for transfers from each dilution.

3. Place dilution volume in a sterile petri dish. Lift the cover of the petri dish just high enough to insert pipet. Let the diluted sample drain from the pipet until delivered. Touch the tip of the pipet once against a dry spot on the petri dish bottom. Remove the pipet without touching it to the dish.

4. Prepare at least two replicate plates for each sample dilution used.

5. After placing test portions for each dilution series on the plates, pour the culture medium and mix carefully. Do not allow more than 20 min to elapse between starting pipeting and pouring plates.

Melting Agar Medium.

1. Melt sterile solid agar medium in boiling water. Discard melted agar that contains precipitate.

2. Maintain melted medium in a water bath between 44°C and 46°C until used. Place a thermometer in water or medium in a separate dilution bottle to monitor medium temperature when pouring agar medium.

Pouring Plates.

1. Limit the number of samples to be plated so no more than 20 min elapse between diluting the first sample and pouring the last plate.

2. Pour 10 to 12 mL melted medium maintained at 44°–46°C into each petri dish by gently lifting the cover just high enough to pour the medium.

3. As each plate is poured, carefully mix the medium with sample by gently rotating the petri dish, first in one direction and then in the opposite direction. Do not allow medium to splash on cover or dish edges.

4. Allow the plates to solidify on a level surface. When the medium is solid, invert the plates and place in the incubator.

5. Incubate at 35°C for 48 hr. Count all colonies on plates promptly after incubation. Report counts as CFU/mL. Alternatively, plates may be held at room temperature (25°C) for 5 to 7 days.

Fecal Coliform and *E. coli*

All positive distribution system total coliform tests for public water supplies must be tested for either fecal coliform or *E. coli*. These procedures are discussed in the *Federal Register*, 40 CFR 141 and 142. The procedures determine the presence or absence of fecal coliform and *E. coli* but do not estimate density.

For any positive MF, MTF, or P–A tests for total coliform, a fecal coliform or *E. coli* analysis is required. For the fecal coliform test, positive total coliform tubes or membranes are used to inoculate the fecal coliform EC medium. The presence or absence of fecal coliform or *E. coli* should be done simultaneously with the confirmation of total coliform in brilliant green bile. Gas production in the inner fermentation tube using EC medium after incubating for 24 hr ± 2 hr at 44.5°C is a positive fecal coliform test. Turbidity with no gas production is a negative test.

An approved regulatory test procedure to confirm the presence of *E. coli* requires the use of EC medium + MUG. The EC medium + MUG is inoculated using presumptive positive total coliform samples from the MF, P–A, or MTF tests. The inoculated EC + MUG medium is examined for fluorescence after 24 hr ± 2 hr of incubation at 44.5° ± 0.5°C.

A second approved *E. coli* procedure requires the transfer of the membrane filter with presumptive positive total colifoms to nutrient agar + MUG, which is examined for fluorescence after 4 hr of incubation at 35°C.

Procedure (EC Media).

1. When using the P–A or MTF technique, shake the positive tubes or bottles vigorously to distribute bacteria uniformly throughout the medium. Use a sterile 3-mm loop or applicator stick to transfer a portion of the positive presumptive lactose or lauryl tryptose broth tubes or bottles to inoculate both the fecal coliform medium and brilliant green bile broth.

   For the MF technique, transfer a small portion of any coliform colony, as many as 10 colonies, with an applicator stick to brilliant green bile. Remove the entire membrane filter from the petri dish with sterile forceps and insert into the tube with EC medium. Curl or quarter the filter using forceps. Because of the limited size of most forceps, a sterile applicator stick may be required to force the filter down into the tube and below the liquid medium level. Be extremely careful not to crack the fermentation tube or cause it to rise when removing the applicator or forceps.

   Another variation in the fecal coliform procedure permits swabbing a membrane filter with a cotton swab to inoculate the EC medium. Use the same swab to inoculate other total coliform confirmation media, such as brilliant green bile broth. Alternatively, laboratories select individual colonies on the membrane filter for transfer to the total coliform confirmation medium, then swab the membrane for transfer to the EC medium. Remove the swab from the EC medium.

2. After inoculating the EC medium tube, gently shake it to ensure adequate mixing. Incubate the EC tube in a water bath or solid block incubation for

24 hr ± 2 hr at 44.5° ± 0.2°C. Standard dry air commercial incubators do not have the sensitivity to maintain a ± 0.2°C temperature setting.

3. After the incubation period, examine the EC tubes for gas production. If gas is produced in any amount, the tube is positive; negative tubes do not have gas produced. Report the results as fecal coliform present or absent.

Quality Control.

When USEPA publishes the mandatory quality control procedures for these techniques, these procedures are expected to closely mimic the MTF procedures. The procedures include but are not limited to

1. Record the water bath or solid block incubator temperature twice a day at least 4 hr apart.

2. Record the lot numbers and date received for ingredients for all media and reagents.

3. Record refrigerator temperature at least daily.

4. Record the dates samples were received, plated, and read.

5. Record what was autoclaved, when and how long.

6. Use a thermometer with gradations of at least 0.2°C in a water bath (0.1° gradations may be needed).

7. Run a positive control of each media lot or batch.

8. Check and record thermometer calibration against a registered thermometer (at least annually).

9. For disposable glassware, reseal the packaging between reuses.

10. Use a nontoxic inoculating loop.

11. Maintain a sufficient water depth in the water bath incubator to immerse tubes to the upper level of the medium.

12. Ensure that the water bath incubator lid completely covers the unit to minimize evaporation losses.

Procedures (*E. coli*).

EC Medium + MUG.

1. Perform a total coliform analysis using the MF, MTF, or P–A test. Use all suspected total coliform samples to inoculate the EC medium + MUG. Use traditional methods, such as a loop, when the MTF or P–A test is used. For the MF test, use a cotton swab over the entire filter area and inoculate both the EC medium + MUG and a total coliform confirmation medium. Remove the swab from the media. A Durham tube is not required in the EC medium + MUG. An alternative for the presumptive positive MF requires sterile forceps to transfer the entire membrane into the EC medium + MUG.

2. The method relies on visual fluorescence under UV light after incubation at 44.5°C for 24 hr to detect *E. coli*. Perform the test for *E. coli* simultaneously with the total coliform confirmation tests.

Nutrient Agar Plus MUG.

1. Transfer an MF with suspected total coliform colonies to nutrient agar supplemented with 100 µg/mL of MUG.

2. The presence of *E. coli* is detected by visual fluorescence under UV light after 4 hr of incubation at 35°C. Before transferring the membrane filter, inoculate total coliform confirmation media.

This page intentionally blank

Appendix **A**

References and Other Sources of Information

REFERENCES

Aieta, E.M., P.V. Roberts, and M. Hernandez. 1984. Determination of Chlorine Dioxide, Chlorine, Chlorite, and Chlorate in Water. *Jour. AWWA,* 76:64.

American Public Health Association, American Water Works Association and Water Environment Federation. *Standard Methods for the Examination of Water and Wastewater.* Latest edition. (Edited by A.D. Eaton, L.S. Clesceri, and A.E. Greenberg.) Washington, D.C.: APHA. (Available from the American Water Works Association.)

American Society for Testing and Materials. June 2001. *Annual Book of ASTM Standards.* Vol. 11.02 on Water. Philadelphia, Pa.: ASTM.

Bates, R.G. 1978. Concept and Determination of pH. In *Treatise on Analytical Chemistry.* Edited by I.M. Kolthoff and P.J. Elving. 1:821. New York: Wiley-Interscience.

Flavor Profile Analysis: Screening and Training of Panelists. 1993. Denver, Colo.: American Water Works Association.

Gordon, G., W.J. Cooper, R.G. Rice, and G.E. Pacey. 1987. *Research Report: Disinfectant Residual Measurement Methods.* Denver, Colo: American Water Works Association.

Hach, C.C. 1972. Understanding Turbidity Measurement. In *Ind. Water Eng.* 9(2):18.

Hach Company. *Water Analysis Handbook.* 1997. Loveland, Colo.: Hach Company.

Jar Testing. 1995. Water Supply Operations Video Series. Denver, Colo.: American Water Works Association.

Lisle, J. 1993. *An Operator's Guide to Bacteriological Testing.* Denver, Colo: American Water Works Association.

Miller, R.J. Tips for the Safe Storage of Laboratory Chemicals. 1985. *Digester/Over the Spillway.* Urbana, Ill.: Illinois Environmental Protection Agency, May–June.

Operational Control of Coagulation and Filtration Processes. Manual M37. 1992. Denver, Colo.: American Water Works Association.

Ozone in Water Treatment—Applications, Operations, and Technology. 1985. Denver, Colo: American Water Works Association.

SDWA Advisor. 2001 Regulatory Update Service. Denver, Colo.: American Water Works Association.

Sequeira, J. Laboratory Procedures. 1992. *Water Treatment Plant Operation*, Vol. 1. K.D. Kerri, Project Director. California State University. Sacramento, Calif.: School of Engineering.

Sokoloff, V.P. 1933. Water of crystallization in total solids of water analysis. In *Ind. Eng. Chem. Anal.* Ed. 5:336.

US Environmental Protection Agency. 1989. Drinking Water; National Primary Drinking Water Regulations; Total Coliforms (including Fecal Coliforms and *E. coli*). *Federal Register.* 54:124:27545 (June 29, 1989).

US Environmental Protection Agency. 1979. *Methods of Chemical Analysis of Water and Wastes*. EPA-600/4-83-020. Cincinnati, Ohio: USEPA.

Utility Chemical Safety. 1994. *Waterworld Review.* Mar/Apr.

Water Quality. 1995. Principles and Practices of Water Supply Operations. Vol. 4. 2nd ed. Denver, Colo.: American Water Works Association.

TECHNICAL ASSISTANCE

AWWA Small Systems Support: 1-800-366-0107

PRODUCT INFORMATION AND, IN SOME CASES, TECHNICAL ASSISTANCE

The American Water Works Association does not endorse or recommend products. The following is not a comprehensive list of manufacturers and suppliers.

Aldrich Chemical Company. *Aldrich Handbook of Fine Chemicals.*
   Phone 800-558-9160
   Fax 800-962-9591
   www.sigma-aldrich.com

Chemetrics, Inc. *Advanced Systems for Water Analysis.*
   Phone 800-356-3072
   Fax 540-788-4856

Fisher Scientific. *The Fisher Catalog.*
   Phone 800-766-7000
   Fax 800-926-1166
   www.fishersci.com

Hach Company. *Products for Analysis.*
   National and International Sales, and Technical Assistance
   Phone 800-227-4224
   Free technical literature on laboratory and portable instruments, turbidimeters, chlorine testing, *Water Analysis Handbook*, 3rd ed., #8374
   www.hach.com

HF Scientific. *HF Scientific.*
   Phone 941-337-2116
   Fax 941-332-7643
   www.hfscientific.com

IDEX Labs. *Colilert™.*
   Phone 800-321-0207
   Fax 207-856-0630
   www.idexx.com

LaMotte Chemical Products Co. *Water Quality Testing Products.*
      Phone 410-778-3100
      Fax 410-778-6394
      www.lamotte.com

Millipore Corp.
      Phone 800-645-5476
      Fax 800-645-5439
      www.millipore.com

Orion Research Inc.
      Phone 800-225-1480
      Fax 978-232-6015
      www.orionres.com

Sigma Chemical Company. *Sigma Chemical Company.*
      Phone 800-521-8956
      Fax 800-325-5052
      www.sigmaus.com

Thomas Scientific. *Thomas Scientific Laboratory Supplies & Reagents.*
      Phone 800-345-2100
      Fax 609-467-3087
      www.thomassci.com

Utility Supply of America. *The USA Blue Book.*
      National and International
      Phone 800-548-1234
      Fax 847-272-8914

USEPA–OGWPW
      Web site (www.epa.gov/ogwdw) and/or the USEPA methods reference of
      (www.epa.gov/safewater/methods/methods.html)

Van Waters and Rogers Scientific Products. *VWR Scientific.*
      Phone 800-932-5000
      Fax 856-467-3336
      www.vwrsp.com

This page intentionally blank

Appendix **B**

Safe Storage of Laboratory Chemicals

The following information is intended to provide guidance for laboratory personnel for safe chemical storage. Appendix B is derived from an article by Robert J. Miller in *Digester/Over the Spillway*, a publication of the Illinois Environmental Protection Agency.

As stated in *Standard Methods for the Examination of Water and Wastewater*, "All laboratory employees must make every effort to protect themselves and their fellow workers." When safety programs are set up, a group effort should be made so everyone involved in laboratory analysis will be trained in laboratory safety procedures. Implementing safe chemical storage procedures can avoid potential hazards in the laboratory.

CHEMICAL STORAGE

Proper chemical storage avoids storing incompatible chemicals together. With the wide variety of chemicals needed for analyses, it is essential to know the specific storage needs for certain chemical groups. Alphabetical storage of chemicals is not always the best approach. Table B-1 lists chemicals that should not be stored together or that have special storage requirements. Laboratories should be equipped with a variety of dedicated chemical storage areas, including an acid cabinet and an explosion-proof refrigerator.

Acids

Keep acids in an acid cabinet. If an acid cabinet is unavailable, store on the lowest shelf. Certain acids, such as nitric, chromic, and acetic acids, are incompatible storage partners. Separate oxidizing agents from organic acids.

Bases

Store bases such as ammonium hydroxide, sodium bicarbonate, and calcium carbonate away from acids. Keep inorganic hydroxide solutions, such as sodium hydroxide, in polyethylene bottles or borosilicate glass bottles with rubber or neoprene stoppers.

Flammable Liquids

Store flammable liquids such as acetone, alcohols, or ethers in safety cans or cabinets or an explosion-proof refrigerator. Sources of heat and sparks must be kept away from this group of chemicals.

Peroxide-Forming Chemicals

Store peroxide-forming chemicals such as ethyl ether, isopropyl ether, and acetaldehyde in airtight containers in a dark, cool place such as an explosion-proof refrigerator. The time factor involved with these chemicals is crucial because explosive peroxides will form over time. Once formed, just turning the cap or a minor jolt may cause an explosion. Date these chemicals both on receiving and opening so storage time can be controlled. The longer the storage period, the greater the amount of peroxides that form. Check to see whether an inhibitor has been added to prevent the formation of peroxides. If not, the chemical should not be kept for more than one year unopened or six months after opening.

Oxidizers

Store oxidizers, including iodine, sodium iodate, potassium peroxide, sodium peroxide, chromic acid, nitric acid, sulfuric acid, perchloric acid, and hydrogen peroxide, in a cool, dry area away from flammable materials and reducing agents. Keep wood and paper products away from this group.

Water-Reactive Chemicals

Keep chemicals that react with water to form toxic substances in a cool, dry area. Water-based fire extinguishers should never be used in this storage area.

Pyrophoric Substances

Chemicals that react with air and ignite are called pyrophoric substances. Examples include fine granules of cadmium, calcium, iron, lead, manganese, and zinc. Store these chemicals in a cool, dry area.

Light-Sensitive Chemicals

Store light-sensitive chemicals, such as ethyl ether, mercuric chloride, mercuric iodide, and sodium iodide, in amber bottles away from direct sunlight.

Toxic Chemicals

Handle toxic compounds, carcinogens, and teratogens (agents that cause birth defects) with extreme care. Labels must notify the user of their danger. Store these chemicals according to their hazardous nature. Avoid inhalation, skin contact, and ingestion. Some of the frequently used chemicals in this group include arsenic compounds, sodium hypochlorite, cadmium compounds, carbon tetrachloride, fluoride compounds, hydrogen peroxide, iodine, mercury, lead, benzene, mercuric compounds, chloroform, phenol, phosphorus, potassium, toluene, sodium, turpentine, and sodium hydroxide.

Media

Store dry media, such as M-Endo broth and MFC broth, in a dessicator in a cool, dry area after opening. Ampouled media, in liquid form, should be refrigerated.

SUMMARY

Review your chemical storage facility and be sure that the following safety measures have been taken:

- Dates are recorded for all chemicals on receiving and opening.
- Bottles show no damage from shipping.
- Storage areas are level and secure.
- Storage areas are free from overcrowding.
- Chemicals are easily seen and within reach.
- Large containers are stored no higher than 2 ft (0.6 m) above the floor.
- Storage areas are well lighted and away from sources of heat and humidity.
- Exit areas are well marked.
- Good housekeeping is evident.
- Safety reminders are posted in specific areas.
- Manufacturers' phone numbers are available if any questions arise.

Table B-1   List of incompatible chemicals

These Chemicals	Should Not Be Stored With or Near These Chemicals
Acetic acid	Chromic acid, nitric acid, hydroxyl compounds, ethylene glycol, perchloric acid, peroxides, permanganates
Acetylene	Chlorine, bromine, copper, fluorine, silver
Alkaline metals, such as other chlorinated hydrocarbons, carbon dioxide, the halogens	Water, carbon tetrachloride, or sodium and potassium, or powdered aluminum, or magnesium
Ammonia, anhydrous	Mercury, chlorine, calcium hypochlorite, iodine, bromine, hydrofluoric acid (anhydrous)
Ammonium nitrate	Acids, powdered meals, flammable liquids, chlorates, nitrites, sulfur, finely divided organic or combustible materials
Carbon, activated	Calcium hypochlorite, all oxidizing agents
Chlorates	Ammonium salts, acids, powdered metals, sulfur, finely divided organic or combustible materials
Chromic acid	Acetic acid, naphthalene, camphor, glycerine, turpentine, alcohol, flammable liquids in general
Chlorine	Ammonia, acetylene, butadiene, butane, methane, propane (or other petroleum gases), hydrogen, sodium carbide, turpentine, benzene, finely divided metals
Copper	Acetylene, hydrogen peroxide
Flammable liquids	Ammonium nitrate, chromic acid, hydrogen peroxide, nitric acid, sodium peroxide, the halogens
Fluorine	Isolate from everything
Hydrocarbons	Fluorine, chlorine, bromine, chromic acid, sodium peroxide
Hydrofluoric acid, anhydrous	Ammonia, aqueous or anhydrous
Hydrogen peroxide	Copper, chromium, iron, most metals or their salts, alcohols, acetone, organic materials, aniline, nitromethane, flammable liquids, combustible materials
Hydrogen sulfide	Fuming nitric acid, oxidizing gases
Mercury	Acetylene, fulminic acid, ammonia, oxalic acid
Nitric acid, concentrated	Acetic acid, aniline, chromic acid, hydrocyanic acid, hydrogen sulfide, flammable liquids, flammable gases
Oxalic acid	Silver, mercury
Potassium permanganate	Glycerin, ethylene glycol, benzaldehyde, sulfuric acid
Silver	Acetylene, oxalic acid, tartaric acid, ammonium compounds
Sulfuric acid	Potassium chlorate, potassium perchlorate, potassium permanganate, or similar compounds with light metals

After R.J. Miller 1985

Appendix **C**

List of Compounds

The following list provides the name of compounds found in this manual and the chemical formula for each.

Name	Formula
acetic acid	CH_3COOH
aluminum hydroxide	$Al(OH)_3$
aluminum sulfate	$Al_2(SO_4)_3$
ammonia	NH_3
ammonium acetate	$NH_4C_2H_3O_2$
ammonium chloride	NH_4Cl
ammonium hydroxide	NH_4OH
ammonium molybdate	$(NH_4)_6Mo_7O_{24}$
ammonium persulfate	$(NH_4)_2S_2O_8$
ammonium sulfate	$(NH_4)SO_4$
barium chloride	$BaCl_2$
barium hydroxide	$Ba(OH)_2$
barium sulfate	$BaSO_4$
bicarbonate	HCO_3^-
boric acid	H_3BO_3
calcium carbonate	$CaCO_3$
calcium hydroxide	$Ca(OH)_2$
carbon dioxide	CO_2
carbonate	CO_3^{2-}
chlorate	ClO_3^-
chlorine dioxide	ClO_2
chlorine (combined) an example of a combined chlorine	NH_2Cl
chlorine (free)	Cl_2

Name	Formula
chlorite	ClO_2^-
chloroplatinic acid	H_2PtCl_6
chromium (VI)	Cr^{6+}
cobaltous chloride	$CoCl_2$
copper sulfate	$CuSO_4$
dichloramine	$NHCl_2$
disodium hydrogen phosphate, *also called* sodium dibasic phosphate	Na_2HPO_4
ferric chloride	$FeCl_3$
ferrous ammonium sulfate	$Fe(NH_4)_2(SO_4)_2$
hydrazine sulfate	$N_2H_4 \cdot H_2SO_4$
hydrochloric acid	HCl
hydrogen peroxide	H_2O_2
hydrogen sulfide	H_2S
hydroxide	OH^-
hypochlorite ion	OCl^-
hypochlorous acid	$HOCl$
magnesium chloride	$MgCl_2$
magnesium oxide	MgO
magnesium sulfate	$MgSO_4$
manganese sulfate	$MnSO_4$
mercuric chloride	$HgCl_2$
mercuric iodide	HgI_2
monochloramine	NH_2Cl
nitrate	NO_3^-
nitric acid	HNO_3
nitrite	NO_2^-
nitrogen trichloride	NCl_3
oxalic acid	$H_2C_2O_4$
phenanthroline	$C_{12}H_8N_2$
phenylarsine oxide	C_6H_5AsO
phosphoric acid	H_3PO_4
potassium bi-iodate	$KH(IO_3)_2$
potassium iodide	KI
potassium chloride	KCl
potassium chloroplatinate	K_2PtCl_6
potassium dihydrogen phosphate (*also called* potassium monobasic phosphate)	KH_2PO_4
potassium nitrate	KNO_3
potassium permanganate	$KMnO_4$
potassium persulfate	$K_2S_2O_8$
silicon dioxide	SiO_2
silver sulfate	Ag_2SO_4

Name	Formula
sodium acetate	$NaC_2H_3O_2$
sodium arsenate	$NaAsO_4$
sodium arsenite	$NaAsO_2$
sodium azide	NaN_3
sodium bicarbonate	$NaHCO_3$
sodium bisulfite	$NaHSO_3$
sodium borate	$Na_2B_4O_7$
sodium carbonate (soda ash)	Na_2CO_3
sodium chloride	$NaCl$
sodium fluoride	NaF
sodium hydroxide	$NaOH$
sodium hypochlorite	$NaOCl$
sodium metasilicate	$NaSiO_3$
sodium nitrite	$NaNO_2$
sodium oxylate	$Na_2C_2O_4$
sodium sulfate	Na_2SO_4
sodium sulfite	Na_2SO_3
sodium tartrate	$Na_2C_4H_4O_6$
sodium thiosulfate	$Na_2S_2O_3$
stannous chloride	$SnCl_2$
sulfamic acid	H_2NSO_3H
sulfanilamide	$C_6H_8N_2O_2S$
sulfur dioxide	SO_2
sulfuric acid	H_2SO_4
trichloramine	NCl_3
zinc sulfate	$ZnSO_4$

This page intentionally blank

Glossary

Acidic The condition of water or soil that contains a sufficient amount of acid substances to lower the pH below 7.0.

Alkali Any of certain soluble salts, principally of sodium, potassium, magnesium, and calcium, that have the property of combining with acids to form neutral salts and may be used in chemical water treatment processes.

Alkaline The condition of water or soil that contains a sufficient amount of alkali substances to raise the pH above 7.0.

Ambient Temperature Temperature of the surrounding air (or other medium). For example, temperature of the room where a gas chlorinator is installed.

Amperometric Titration A means of measuring concentrations of certain substances in water (such as strong oxidizers) based on the electric current that flows during a chemical reaction. See *Titrate*. Frequently used to measure disinfectant residuals.

Aseptic Sterile. Free from all living organisms, including the living germs of disease, fermentation, or putrefaction.

Bacteria Single-celled, microscopic, living organisms.

Blank A bottle that contains only dilution water or distilled water; the sample being tested is not added to the bottle containing the blank. Tests are frequently run on a sample and a blank to compare differences.

Buffer A solution or liquid whose chemical makeup neutralizes acids or bases without a large change in pH.

Buffer Capacity The ability of a solution or liquid to neutralize acids or bases. This is a measure of the capacity of water for offering a resistance to changes in pH.

Calcium Carbonate ($CaCO_3$) Equivalent An expression of the concentration of specified constituents in water in terms of their equivalent value to calcium carbonate. For example, the hardness in water that is caused by calcium, magnesium, and other ions is usually described as calcium carbonate equivalent.

Carcinogen Any substance that tends to produce cancer in an organism.

Chlorine-Demand-Free Water Prepare chlorine-demand-free water from good-quality distilled or deionized water by adding sufficient chlorine to give 5 mg/L free chlorine. After standing 2 days, this solution should contain at least 2 mg/L free chlorine; if not, discard and obtain better-quality water. Remove remaining free chlorine by placing container in sunlight or irradiating with an ultraviolet lamp. After several hours, take sample, add KI, and measure total chlorine with a colorimetric method, using a Nessler tube to increase sensitivity. Do not use in the water before last trace of free and combined chlorine has been removed.

Colorimetric Measurement A means of measuring unknown chemical concentrations in water by measuring a sample's color intensity. The specific color of the sample, developed by adding chemical reagents, is measured with a photo-electric colorimeter or is compared with color standards using, or corresponding with, known concentrations of the chemical. Colorimetric measurements may be made with color comparison tubes, colorimeters, spectrophotometers, and the older photometers.

Composite Samples   A collection of individual samples obtained at regular intervals, usually every one or two hours during a 24-hr time period. Each individual sample is combined with the others in proportion to the rate of flow when the sample was collected.

Compound   A pure substance composed of two or more elements whose composition is constant. NaCl (sodium chloride, also known as table salt) is a compound.

Desiccator   A closed container into which heated weighing or drying dishes are placed to cool in a dry environment in preparation for weighing. Desiccators contain a substance that absorbs moisture and keeps the relative humidity near zero so the dish or sample will not gain weight from absorbed moisture from the air.

Disinfection   The process designed to kill most microorganisms in water, including pathogenic bacteria.

DPD   A method of measuring the chlorine residual in water. The residual may be determined by either titrating to an end point or comparing a developed color with color standards. DPD is an abbreviation for *N, N*-diethyl-*p*-phenylene-diamine.

Element   A substance that cannot be separated into its constituent parts and still retain its chemical identity.

End Point   The completion of a desired chemical reaction. Samples are titrated to the end point. An end point may be detected by the use of an electronic device such as a pH meter or a color change.

Grab Sample   A single sample of water collected at a particular time and place. The sample represents the composition of the water only at that time and place.

Gravimetric   A means of measuring constituents in a sample based on weight. Total and suspended solids are examples.

Indicator (Chemical)   A substance that gives a visible change, at a desired point in chemical reaction. Indicators usually are color based and demonstrate a specific end point to a chemical reaction.

Inorganic   Material such as sand, salt, iron, calcium salts, and other mineral materials. Inorganic substances are of mineral origin; organic substances are usually of animal or plant origin.

Meniscus   The curved top of a column of liquid (water, oil, mercury) in a small tube. When the liquid wets the sides of the container (as with water), the curve forms a valley. When the confining sides are not wetted (as with mercury), the curve forms a hill or upward bulge.

Milligrams per Liter (mg/L)   A measure of the concentration by weight of a substance per unit volume. For practical purposes, one mg/L of a substance in fresh water is equal to one part per million parts (ppm).

Mole   The gram molecular weight of a substance, usually expressed in grams.

Molecule   The smallest division of a compound that still retains or exhibits all the properties of the substance.

MPN   The most probable number of bacteria per unit volume of sample water. Expressed as an estimate of the density or population of organisms per 100 mL of sample water.

_N_ or Normal   A normal solution contains one gram equivalent weight of compound per liter of solution.

Nephelometric   A means of measuring turbidity by using an instrument called a nephelometer, which passes light through a sample and the amount of light deflected (usually at a 90-degree angle) is then measured.

OSHA   The Williams-Steiger Occupational Safety and Health Act of 1970 (OSHA) is a federal law designed to protect the health and safety of industrial workers including water supply system and treatment plant workers. OSHA also refers to the federal and state agencies that administer the OSHA regulations.

Organic   Substances that come from animal or plant sources. Organic substances always contain carbon. Inorganic materials are chemical substances of mineral origin. Also see _Inorganic_.

Organism   Any form of animal or plant life. Also see _Bacteria_.

Oxidation   The addition of oxygen, removal of hydrogen, or the removal of electrons from an element or compound. In the environment, organic matter is oxidized to more stable substances.

Oxidation-Reduction Potential (ORP)   The electrical potential required to transfer electrons from one compound or element (the oxidant) to another compound or element (the reductant); used as a qualitative measure of the state of oxidation in water treatment systems.

Parts Per Million (ppm)   A measurement of concentration by weight or volume. For practical purposes, this term is equivalent to milligrams per liter (mg/L), which is the preferred term.

Pathogenic Organisms   Organisms, including bacteria, fungi, viruses or protozoans, capable of causing diseases. There are many types of bacteria that _do not_ cause disease. These organisms are nonpathogenic or saprophytic forms.

Pathogens   Disease-causing organisms.

Percent Saturation   The amount of a substance that is dissolved in a solution compared with the amount that could be dissolved in the solution expressed as a percent.

pH   An expression of the intensity of the basic or acidic condition of a liquid. The pH may range from 0 to 14, where 0 is most acidic, 14 most basic, and 7 neutral. Natural waters usually have a pH between 6.5 and 8.5.

Potable Water   Water that does not contain objectionable pollution, contamination, minerals, or infective agents and is considered satisfactory for drinking.

Precipitate   _When used as a noun_, an insoluble, finely divided substance that is a product of a chemical reaction within a liquid. _When used as a verb_, the separation from solution of an insoluble substance.

Protozoans   Single celled organisms which may be found in untreated water supplies and can be pathogenic. _Cryptosporidia_ and _Giardia_ are pathogenic protozoans.

Reagent   A pure chemical substance that is used to make new products or is used in chemical tests to measure, detect, or examine other substances.

Representative Sample   A sample portion of material or water that is as nearly identical in content and consistency as possible to that of a larger body of material or water that is sampled.

Solute   The substance which is dissolved in a solution.

Solution   A liquid mixture of dissolved substances. In a solution all the separate parts cannot be seen.

Solvent   A substance, usually liquid, that dissolves or can dissolve another substance.

Standard Solution   A solution in which the exact concentration of a chemical or compound is known.

Standardize   To compare with a standard. To determine the exact strength of a solution by comparing it with a standard of known strength. This information is used to adjust the strength by adding more water or more of the substance dissolved. To set up an instrument or device to read a standard.

Sterilization   The removal or destruction of all microorganisms, including pathogenic and other bacteria, vegetative forms, and spores.

Supernatant   Liquid removed from settled sludge. Supernatant commonly refers to the liquid between the sludge on the bottom and the water surface of a basin or container.

Surfactant   Abbreviation for surface-active agent. The active agent in detergents that possess a high cleaning ability.

TFE   A fluorinated polymer, polytetrafluoroethylene, used for labware.

Titrate   The process of adding the chemical reagent in increments until completion of the reaction, as signaled by the end point.

Turbidity Units (ntu)   A measure of the cloudiness of water. Turbidity measured by a nephelometer is reported as ntu (nephelometric turbidity units).

Volatile   A substance that can be evaporated or changed to a vapor at relatively low temperatures. Volatile also refers to materials lost (including most organic matter) upon ignition in a muffle furnace for 60 min at 550°C.

Volatile Acids   Fatty acids produced during digestion. They are soluble in water and can be steam-distilled at atmospheric pressure. Also called organic acids, volatile acids are commonly reported as equivalent to acetic acid.

Volatile Liquids   Liquids that easily vaporize or evaporate at room temperature.

Volatile Solids   Those solids in water or other liquids that are lost on ignition of the dry solids at 550°C.

Volumetric   A measurement based on the volume of some factor. Volumetric titration is a means of measuring unknown concentrations of constituents in a sample by determining the volume of titrant or liquid reagent needed to complete particular reactions. Volumetric glassware is used to make precise dilutions.

Index

NOTE: *f.* indicates a figure; *t.* indicates a table.

This page intentionally blank

AWWA Manuals

M1, *Water Rates*, Fifth Edition, 2000, #30001PA

M2, *Instrumentation and Control*, Third Edition, 2001, #30002PA

M3, *Safety Practices for Water Utilities*, Fifth Edition, 1990, #30003PA

M4, *Water Fluoridation Principles and Practices*, Fourth Edition, 1995, #30004PA

M5, *Water Utility Management Practices*, First Edition, 1980, #30005PA

M6, *Water Meters—Selection, Installation, Testing, and Maintenance*, Second Edition, 1999, #30006PA

M7, *Problem Organisms in Water: Identification and Treatment*, Second Edition, 1995, #30007PA

M9, *Concrete Pressure Pipe*, Second Edition, 1995, #30009PA

M11, *Steel Pipe—A Guide for Design and Installation*, Fourth Edition, 1989, #30011PA

M12, *Simplified Procedures for Water Examination*, Second Edition, 1997, #30012PA

M14, *Recommended Practice for Backflow Prevention and Cross-Connection Control*, Second Edition, 1990, #30014PA

M17, *Installation, Field Testing, and Maintenance of Fire Hydrants*, Third Edition, 1989, #30017PA

M19, *Emergency Planning for Water Utility Management*, Fourth Edition, 2001, #30019PA

M20, *Water Chlorination Principles and Practices*, First Edition, 1973, #30020PA

M21, *Groundwater*, Second Edition, 1989, #30021PA

M22, *Sizing Water Service Lines and Meters*, First Edition, 1975, #30022PA

M23, *PVC Pipe—Design and Installation*, First Edition, 1980, #30023PA

M24, *Dual Water Systems*, Second Edition, 1994, #30024PA

M25, *Flexible-Membrane Covers and Linings for Potable-Water Reservoirs*, Second Edition, 1996, #30025PA

M26, *Water Rates and Related Charges*, Second Edition, 1996, #30026PA

M27, *External Corrosion—Introduction to Chemistry and Control*, First Edition, 1987, #30027PA

M28, *Cleaning and Lining Water Mains*, First Edition, 1987, #30028PA

M29, *Water Utility Capital Financing*, Second Edition, 1998, #30029PA

M30, *Precoat Filtration*, Second Edition, 1995, #30030PA

M31, *Distribution System Requirements for Fire Protection*, Second Edition, 1992, #30031PA

M32, *Distribution Network Analysis for Water Utilities*, First Edition, 1989, #30032PA

M33, *Flowmeters in Water Supply*, Second Edition, 1997, #30033PA

M34, *Water Rate Structures and Pricing*, Second Edition, 1999, #30034PA

M35, *Revenue Requirements*, First Edition, 1990, #30035PA

M36, *Water Audits and Leak Detection*, Second Edition, 1999, #30036PA

M37, *Operational Control of Coagulation and Filtration Processes*, First Edition, 1992, #30037PA

M38, *Electrodialysis and Electrodialysis Reversal*, First Edition, 1995, #30038PA

M41, *Ductile-Iron Pipe and Fittings*, First Edition, 1996, #30041PA

M42, *Steel Water-Storage Tanks*, First Edition, 1998, #30042PA

M44, *Distribution Valves: Selection, Installation, Field Testing, and Maintenance*, First Edition, 1996, #30044PA

M45, *Fiberglass Pipe Design*, First Edition, 1996, #30045PA

M46, *Reverse Osmosis and Nanofiltration*, First Edition, 1999, #30046PA

M47, *Construction Contract Administration*, First Edition, 1996, #30047PA

M48, *Waterborne Pathogens*, First Edition, 1999, #30048PA

M50, *Water Resources Planning*, First Edition, 2001, #30050PA

M51, *Air-Release, Air/Vacuum, and Combination Air Valves*, First Edition, 2001, #30051PA

To order any of these manuals or other AWWA publications, call the Bookstore toll-free at 1-(800)-926-7337.

This page intentionally blank

Preparing Common Types of Desk Reagents

ACID SOLUTIONS

Prepare the following reagents by cautiously adding the required amount of concentrated acid, with mixing, to the designated volume of distilled water. Dilute to 1,000 mL and mix thoroughly.

See Table A for preparing hydrochloric acid (HCl), sulfuric acid (H_2SO_4), and nitric acid (HNO_3) solutions.

ALKALINE SOLUTIONS

Stock sodium hydroxide, 15N (for preparing 6N, 1N, and 0.1N solutions). Cautiously dissolve 625 g solid sodium hydroxide (NaOH) in 800 mL distilled water to form 1 L solution. Remove the sodium carbonate (Na_2CO_3) precipitate by keeping the solution at the boiling point for a few hours in a hot water bath or by letting particles settle for at least 48 h in an alkali-resistant container (wax-lined or polyethylene) protected from atmospheric carbon dioxide (CO_2) with a soda lime tube. Use the supernate for preparing dilute solutions listed in Table B.

Alternatively, prepare dilute solutions by dissolving the weight of solid sodium hydroxide indicated in Table B in CO_2^- free distilled water and diluting to 1,000 mL.

Store sodium hydroxide solutions in polyethylene (rigid, heavy-type) bottles with polyethylene screw caps, paraffin-coated bottles with rubber or neoprene stoppers, or borosilicate-glass bottles with rubber or neoprene stoppers. Check solutions periodically. Protect them by attaching a tube of carbon dioxide-absorbing granular material such as soda lime or a commercially available carbon dioxide-removing agent.[1] Use at least 70-cm rubber tubing to minimize the vapor diffusion from the bottle. Replace the absorption tube before it becomes exhausted. Withdraw the solution with a siphon to avoid opening the bottle.

Ammonium hydroxide solutions. Prepare 5N solutions by diluting 333 mL, 3N by diluting 200 mL, and 0.2N ammonium hydroxide (NH_4OH) solutions by diluting 13 mL of concentrated reagent (sp gr 0.90, 29.0 percent, 15N) to 1,000 mL with distilled water.

INDICATOR SOLUTIONS

Phenophthalein indicator solution: Use either the aqueous (1) or alcoholic (2) solution.

1. Dissolve 5 g phenolphthalein disodium salt in distilled water and dilute to 1 L.

2. Dissolve 5 g phenolphthalein in 500 mL 95 percent ethyl or isopropyl alcohol and add 500 mL distilled water.

If necessary, add 0.02N sodium hydroxide drop by drop until a faint pink color appears in solution (1) or (2).

Methyl orange indicator solution. Dissolve 500 mg methyl orange powder in distilled water and dilute to 1 L.

Table B   Preparing uniform sodium hydroxide solutions

Normality of NaOH Solution	Required Weight of NaOH to Prepare 1,000 mL Solution	Required Volume of 15N NaOH to Prepare 1,000 mL Solution
	g	mL
6	240	400
1	40	67
0.1	4	6.7

Table A   Preparing uniform acid solutions*

Desired Component	Hydrochloric Acid (HCl)	Sulfuric Acid (H_2SO_4)	Nitric Acid (HNO_3)
Specific gravity (20/4° C) of ACS-grade concentrated acid	1.174–1.189	1.834–1.836	1.409–1.418
Percent of active ingredient in concentrated reagent	36–37	96–98	69–70
Normality of concentrated reagent	11–12	36	15–16
Volume (mL) of concentrated reagent to prepare 1 L:			
18N solution	—	500 (1 + 1)[†]	—
6N solution	500 (1 + 1)[†]	167 (1 + 5)[†]	380
1N solution	83 (1 + 11)[†]	28	64
0.1N solution	8.3	2.8	6.4
Volume (mL) of 6N reagent to prepare 1 L of 0.1N solution	17	17	17
Volume (mL) of 1N reagent to prepare 1 L of 0.02N solution	20	20	20

*All values are approximate.

†The $a + b$ system of specifying preparatory volumes means that a volumes of the concentrated reagent are diluted with b volumes of distilled water to form the required solution.

[1] Ascarite II,® Arthur H. Thomas Co.; or equivalent.

American Water Works Association

Periodic Table of the Elements

Legend:

Atomic number	1	1.0079	Atomic weight
Name	Hydrogen		
Melting point (°C)	−259.14	−252.87	Boiling point
Symbol	**H**		

Each cell below lists: atomic number, atomic weight (top right); name; melting point, boiling point (°C); symbol.

I-A	II-A	III-B	IV-B	V-B	VI-B	VII-B	VIII	VIII	VIII	I-B	II-B	III-A	IV-A	V-A	VI-A	VII-A	0
1 1.0079 Hydrogen −259.14 −252.87 **H**																	2 4.0026 Helium <−272.2 −268.93 **He**
3 6.914 Lithium 180.54 1347 **Li**	4 9.0122 Beryllium 1278 2970 **Be**											5 10.811 Boron 2079 2550 **B**	6 12.011 Carbon 3367 4827 **C**	7 14.007 Nitrogen −209.86 −195.8 **N**	8 15.999 Oxygen −218.4 −182.96 **O**	9 18.998 Fluorine −219.62 −188.14 **F**	10 20.179 Neon −248.67 −246.05 **Ne**
11 22.990 Sodium 97.81 882.9 **Na**	12 24.305 Magnesium 648.8 1090 **Mg**											13 26.982 Aluminum 660.37 2467 **Al**	14 28.086 Silicon 1410 2355 **Si**	15 30.974 Phosphorus 44.1 280 **P**	16 32.066 Sulfur 112.8 444.67 **S**	17 35.453 Chlorine −100.98 −34.6 **Cl**	18 39.948 Argon −189.2 −185.7 **Ar**
19 39.098 Potassium 63.65 774 **K**	20 40.078 Calcium 839 1484 **Ca**	21 44.956 Scandium 1541 2831 **Sc**	22 47.867 Titanium 1660 3287 **Ti**	23 50.942 Vanadium 1890 3380 **V**	24 51.996 Chromium 1857 2672 **Cr**	25 54.938 Manganese 1244 1962 **Mn**	26 55.845 Iron 1535 2750 **Fe**	27 58.933 Cobalt 1495 2870 **Co**	28 58.693 Nickel 1453 2743 **Ni**	29 63.546 Copper 1083 2567 **Cu**	30 65.39 Zinc 419.6 907 **Zn**	31 69.723 Gallium 29.78 2403 **Ga**	32 72.61 Germanium 937.4 2830 **Ge**	33 74.922 Arsenic 817 613 **As**	34 78.96 Selenium 217 684.9 **Se**	35 79.904 Bromine −7.2 58.78 **Br**	36 83.80 Krypton −156.6 −152.3 **Kr**
37 85.468 Rubidium 38.89 688 **Rb**	38 87.62 Strontium 769 1384 **Sr**	39 88.906 Yttrium 1522 3338 **Y**	40 91.224 Zirconium 1852 4377 **Zr**	41 92.906 Niobium 2468 4742 **Nb**	42 95.94 Molybdenum 2617 4612 **Mo**	43 *98.906 Technetium 2172 4877 **Tc**	44 101.07 Ruthenium 2310 3900 **Ru**	45 102.91 Rhodium 1966 3727 **Rh**	46 106.42 Palladium 1552 3140 **Pd**	47 107.87 Silver 961.9 2212 **Ag**	48 112.41 Cadmium 320.9 765 **Cd**	49 114.82 Indium 156.6 2080 **In**	50 118.71 Tin 232.0 2270 **Sn**	51 121.76 Antimony 630.7 1750 **Sb**	52 127.60 Tellurium 449.5 990 **Te**	53 126.90 Iodine 113.5 184.4 **I**	54 131.29 Xenon −111.9 −107.1 **Xe**
55 132.91 Cesium 28.40 678.4 **Cs**	56 137.33 Barium 725 1640 **Ba**	57 138.91 Lanthanum 921 3457 **La**	72 178.49 Hafnium 2227 4602 **Hf**	73 180.95 Tantalum 2996 5425 **Ta**	74 183.85 Tungsten 3410 5660 **W**	75 186.2 Rhenium 3180 5627 **Re**	76 190.23 Osmium 3045 5027 **Os**	77 192.22 Iridium 2410 4130 **Ir**	78 195.08 Platinum 1772 3827 **Pt**	79 196.97 Gold 1064 2807 **Au**	80 200.59 Mercury −38.84 356.6 **Hg**	81 204.38 Thallium 303.5 1457 **Tl**	82 207.2 Lead 327.5 1740 **Pb**	83 208.98 Bismuth 271.3 1560 **Bi**	84 *208.98 Polonium 254 962 **Po**	85 *209.99 Astatine 302 337 **At**	86 *222.02 Radon −71 −61.8 **Rn**
87 *223.02 Francium 27 677 **Fr**	88 *226.03 Radium 700 1140 **Ra**	89 *227.03 Actinium 1050 3200 **Ac**	104 (Unq)	105 (Unp)	106 (Unh)	107 (Uns)	108 (Uno)	109 (Unn)									

Lanthanide series:

58 140.12 Cerium 799 3426 **Ce**	59 140.91 Praseodymium 931 3512 **Pr**	60 144.24 Neodymium 1021 3068 **Nd**	61 *146.92 Promethium 1168 2460 **Pm**	62 150.36 Samarium 1077 1791 **Sm**	63 151.97 Europium 822 1597 **Eu**	64 157.25 Gadolinium 1313 3266 **Gd**	65 158.93 Terbium 1356 3123 **Tb**	66 162.50 Dysprosium 1412 2562 **Dy**	67 164.93 Holmium 1474 2695 **Ho**	68 167.26 Erbium 1497 2900 **Er**	69 168.93 Thulium 1545 1947 **Tm**	70 173.04 Ytterbium 819 1194 **Yb**	71 174.97 Lutetium 1663 3395 **Lu**

Actinide series:

90 *227.03 Thorium 1750 4790 **Th**	91 *231.04 Protactinium 1600 **Pa**	92 *238.03 Uranium 1132 3818 **U**	93 *237.05 Neptunium 640 3902 **Np**	94 *244.06 Plutonium 641 3232 **Pu**	95 *243.06 Americium 994 2607 **Am**	96 *247.07 Curium 1340 **Cm**	97 *247.07 Berkelium **Bk**	98 *251.08 Californium **Cf**	99 *252.08 Einsteinium **Es**	100 *257.10 Fermium **Fm**	101 *258.10 Mendelevium **Md**	102 *259.10 Nobelium **No**	103 *260.10 (Lawrencium) **Lr**

*Most Stable Isotope

To order additional copies of a laminated chart, please call AWWA 1-800-926-7337 and ask for catalog number 30012A.

Courtesy of Fisher Scientific, Inc., Pittsburg, Penn.

American Water Works Association

1P-1M-30012A-11/97